INFORMATION TECHNOLOGY IN MANUFACTURING PROCESSES

Case studies in technological change

edited by
Graham Winch

rossendale

500 (
150]
LON

ISBI

CONTENTS

PREFACE

When talk of the 'microelectronics revolution' first hit the headlines in the late seventies, a number of research projects were funded to examine its impact on various aspects of our lives. Many of these projects are now coming to fruition, and in December 1982 a conference was held at County Hall, London, funded by IT 82 Year to disseminate these findings to a wider audience. Those attending the conference were drawn from the trade unions, management, and others concerned with information technology in the workplace. This book reports the proceedings of this conference.

I should like to thank my colleagues at Imperial College for their support on the conference and this book, and David Kimpton of the Greater London Council for the efficient organisation of the conference.

Graham Winch
Conwy,
July 1983.

PREFACE

Introduction
New technologies, old problems

Graham Winch

THE AUTOMATION DEBATE

In manufacturing industry, information technology (IT) can be utilised in three interdependent ways - business information such as accounts can be computerised, new or improved products can be developed, and it can be applied to the control of manufacturing processes. This book is about the third development. In this sense it is a contribution to the debate about automation that has been ebbing and flowing since at least World War 2.

The word automation was coined in 1946 by a Ford executive to describe mechanised work handling between transfer machines for engine block manufacture(1), but by 1956 the word possessed myriad meanings(2). These ranged from being the motor of the second industrial revolution, to just the natural development of existing tendencies of mechanisation and scientific management(3).

The debate on the social consequences of automation flourished, too. Walter S. Buckingham argued that

> 'if properly understood and applied, developed and controlled, automation, together with atomic energy, may provide the means to eliminate poverty for the first time in the world'(4).

In Britain R. H. MacMillan espoused the hope

> 'that this new branch of technology may eventually enable us to lift the curse of Adam from the shoulders of man, for machines could indeed become men's slaves rather than their masters, now that practical techniques have been devised for controlling them automatically'.(5)

More pessimistically, Norbert Wiener, the theorist of cybernetics, argued 'that the first industrial revolution, the revolution of 'the dark satanic mills', was the devaluation of the human arm by the competition of machinery. There is no rate of pay at which a United States pick-and-shovel laborer can live which is low enough to compete with the work of a steam shovel as an excavator. The modern industrial revolution is similarly bound to devalue the human brain at least in its simpler and more routine decisions. Of course, just as the skilled carpenter, the skilled mechanic, the skilled dressmaker have in some degree survived the first industrial revolution, so the skilled scientist and the skilled administrator may survive the second. However, taking the second revolution as accomplished, the average human being of mediocre attainments or less has nothing to sell that it is worth anyone's money to buy'.(6)

The social consequences of this development were satirised by Kurt Vonnegut in the novel Player Piano in which he visualised technocratic organisation men ruling a divided society.

John Diebold, the chief protagonist of the automated factory, effectively demolished the pessimistic view, while cautioning against a view of a gilded future around the corner(7). Diebold perhaps was most acute when he argued that

'We are beginning to look at our industrial processes as complete, integrated systems, from the introduction of the raw material until the completion of the final product... One way of defining automation is to say that it is a means of organizing or controlling production processes to achieve optimum use of all production resources - mechanical, material, and human... Fundamentally, I think automation means an optimization of our business and industrial activities'.(8)

While most of the contributions to the debate seemed to see automation technology as something which had been created independently of the dymanics of capitalist society, and merely debated whether it would be used or abused, Diebold implicitly recognised that it is a technology created and used by management for management.

This debate stimulated a large amount of sociological research, ranging from detailed studies of the changing nature of work(9) to analyses of the implications for class structure and action(10). Perhaps the most thorough-going analysis of the implications of automation at the level of the individual plant from the fifties is that of James Bright(11). Through a set of thirteen case studies, he covered most of the issues that have been raised in the contemporary debate. He found that the management of innovation placed major strains on managerial resources, and noted the tendency for projects to overrun both schedules and budgets. Benefits for management came in the form of reductions in lead times and inventory, and increases in productivity and quality. More negatively, he found significant decreases in production flexibility which meant that the sales function tended to be driven by production, and an increase in the size of the maintenance function. The skills of production workers tended to be reduced, while those of indirect workers tended to increase, but few industrial relations problems emerged. Bright's work remains the most comprehensive analysis of automation in manufacturing industry to date.

The debate seemed to lose momentum in the mid sixties, and attention turned more towards computerisation in the office. This may have been because diffusion of the major technical advances of the period was largely complete. In other words, those factories which could be automated at the existing state of the art had been so done. It is worth remembering how limited the possibilities were. The focus of the fifties debate was dedicated automation which depended upon high through-puts of homogeneous products. The state of the art could not handle variability or complexity in the product(12). The bulk of industrial production is either complex, variable, or both, and so the opportunities for automation were non-existant. The remarkable cost and performance curves of microelectronics(13) have allowed the development of cheap programmable automation in the seventies, and allowed new areas of application. Production in small batches can now be automated economically, and the ability to store and manipulate cheaply large amounts of data has allowed the application of IT in the design and production planning functions. These new possibilities for automation, coupled with the wider potential of telematics(14) have opened up the debate again.

From the examples given above, it will be clear that the terms of the contemporary IT debate had been set by the early fifties. Indeed, as the Council for Science and Society have pointed out, the debate has many similarities with that which took place during the industrial revolution(15). The same polarities of apocalyptic vision and manic enthusiasm have emerged as in the early fifties, and talk is of the second industrial revolution for the second time. Stimulated by the debate, a body of research is again building up as to what is actually happening in factories as IT makes its mark. This book presents a set of case studies of the implications of programmable automation. The findings will put many aspects of the IT debate in a new light, while showing that the key issues are much the same as those that Bright analysed for dedicated automation.

THE DIFFUSION OF INFORMATION TECHNOLOGY

Before turning to the implications of information technology in manufacturing processes, it is worth assessing how much of this new technology is actually in use. According to the Policy Studies Institute (PSI) Survey of manufacturing establishments employing more than 20 people, some 22.5% of establishments covering 42.9% of employees in manufacturing are using or plan to use some form of microelectronics based technology in their production processes. Use is highest in the electrical and instrument engineering industries, followed by the food and drink industry, and by far the lowest is clothing and leather manufacture. Process applications tend to be concentrated in the southern part of the country, particularly the south west; in large establishments; and in foreign-owned companies(16). Although the data is not fully analysed for intercorrelations, the limited analysis performed suggests that these are partially independent factors. In other words, each may be an element in explaining the pattern of diffusion of IT. A survey of the mechanical engineering industry broadly confirms this view, although the East and West Midlands, and Yorkshire and Humberside were found to be relatively advanced in the adoption of new process technologies(17).

However, such a national overview tells us nothing about the sort or scale of process application. It cannot distinguish between - to take two examples from one plant in the car industry - £8,000 spent on a microcomputer for monitoring a transfer line and £80m on the automation of body assembly. It is necessary to assess the situation technology by technology - unfortunately, very few figures are available.

The counting of the robot population has acquired an almost obsessive air, but the numbers in the UK stood at 1,200 at the end of 1982, against 80 in 1977(18). The present population is about one third of that in West Germany but larger than that in the similarly sized economies of France and Italy. The UK similarly lags behind W. Germany in the use of Programmable Logic Controllers(19). The diffusion of numerically controlled (NC) machines tools is a good guide to the overall extent of diffusion, because of the widespread applicability of NC technology. In 1982, 2.96% of the machine tool stock was NC in the metal manufacturing and working industries, in comparison to 0.96% in 1976, and around one quarter of engineering establishments had some form of NC (20). UK industry has long had a relatively high concentration of NC, a fact usually explained by the relatively large size of the British aerospace industry(21), although evidence on sales of NC machine tools suggests that diffusion is less rapid in the UK than in West Germany(22). No figures are available for the diffusion of computer aided design (CAD), but because of the relative importance of the electronics and aerospace industries in Britain, it would be expected that it, too, was relatively well diffused in comparison to other European countries, but not to the USA or Japan.

It should be noted that this installed capacity accounts for a higher percentage of output than it does of the total stock of production technology for a number of reasons. Firstly, the concern to pay back on an expensive piece of equipment will encourage relatively high utilisation rates. Secondly, a high level of economic activity encourages innovation in both extra and replacement capacity. Thirdly, the quality benefits of information technology will encourage its use on high value products. Finally, improved cost competitiveness should lead to increased market shares at the expense of non-innovators. These effects become stronger as firms move up the learning curve. Thus the importance of installed information technology in production processes is greater than the bald diffusion figures would indicate. The rate of diffusion is also increasing rapidly, despite the effects of the recession. The PSI survey found that two thirds of reported applications had been installed in 1979 or later, and two thirds of robots have been delivered since 1980.

THE IMPLICATIONS OF INFORMATION TECHNOLOGY

The studies reported in this book examine various aspects of the implications of this pattern of diffusion for manufacturing organisations. The studies are based upon empirical research in sectors which were largely immune from the wave of automation in the fifties. There are important developments taking place in the continuous process industries based upon information technology leading to the implementation of networked control systems(23), but organisationally, these are refinements to what is already a highly automated technology. In the car industry, body assembly and the paint shops are now being automated, leaving only final assembly as a labour intensive process. These developments have been analysed elsewhere(24), but again they are extra elements in industries where automation is already well established. If there is a distinctive 'impact' of IT, it will be strongest in those industries with little experience of automation.

The first contribution by John Bessant summarises his work at the Technology Policy Unit over the last few years. He starts by pointing out that innovation in manufacturing processes is more evolutionary than revolutionary, and calls this continual modification _manufacturing innovation_. After examining the motives for innovation, he moves on to discuss the behavioural implications of manufacturing innovation. The topics of technology awareness, manpower planning, industrial relations, and attitudes to technological change are covered. Finally, in discussing the implications for work organisation of IT, he introduces the concept of a _design space_ as a framework for the analysis of the strategic choices open to innovators in industry. This contribution raises the issues which will be taken up in more detail by the following chapters.

Computer aided design (CAD) is one of the major new technologies of the seventies, and Erik Arnold makes an important contribution to debates about industrial strategy through a discussion of the role of the United States Air Force (USAF) in both its development and and diffusion. He contrasts this story with the paucity of arrangements in the UK. Turning to the adoption of CAD by firms in the British electronics and engineering industries, he finds that the key problem of utilisation is the lack of adequate managerial skills. British managers are very unsure how to evaluate CAD systems, and to use them in the most effective way once they are installed. In conclusion he argues that awareness schemes such as IT 82 can only have a limited effect while the underlying problems remain untouched.

The anthropomorphic nature of robots has earned them more than their fair share of attention, argues James Fleck, and they should be seen as one of a range of new technologies to be utilised as part of manufacturing systems. After outlining the myriad choice confronting aspiring robot users, he goes on to examine the contexts which favour their productive use through a comparison of the characteristics of successful and unsuccessful users. Again, the problems of the lack of experience amongst management are raised, as well as the problems of training workers, particularly maintenance workers. In a provocative conclusion, James Fleck quotes a report which was written nearly 30 years ago which pinpoints education and training as the key factors in the implementation of automation, and asks why nothing was done.

A number of commentators have argued the importance of assessing IT as a system technology, rather than a set of disparate innovations. The linking of CAD with computer aided manufacturing (CAM) systems to form CAD/CAM systems which can take product information right from the conception stage to manufacturing instructions shows the integrating potential that such technologies contain. Graham Winch analyses the sorts of conditions under which such integration is likely to generate organisational stress, and hence pressures for changes in organisation structure. He goes on to suggest that matrix type structures might be a solution to such stresses in certain contexts. Finally, he argues for much greater theoretical awareness in the analysis of the implications of IT in organisations, and indicates that there is a considerable body of literature upon which to draw.

David Buchanan provides more evidence of the range of choices open to innovators with IT. He first analyses the differing motives of senior, middle and junior management in innovation and suggests ways in which the control aspirations of management immediately associated with the production process can come into conflict with the more strategic considerations of senior management. Through examining two examples of process control innovation, he shows the way in which poor job design can not only rob the workers of their skills, but is also counter-productive in the long term for management. He concludes by arguing that the work organisation aspects of innovating with IT have not been given the attention they warrant.

The importance of the context of IT innovation is clearly brought out in the Anglo-German comparison of the use of numerically controlled (NC and CNC) machine tools by Ian Nicholas, Malcolm Warner, Arndt Sorge and Gert Hartmann. They argue that batch size and plant size are crucial variables in whether the NC operator retains the programming function, or whether it is performed by part programmers, thereby deskilling the operator. A third, and possibly more important, variable they point to is the national traditions of organisation structure and training. The less hierarchical organisations and higher skilled workforce in West Germany have encourged operator programming in comparison the the specialist part programming in Britain. In conclusion they argue that the trend towards smaller batch size and greater component complexity mean that a shift towards greater utilisation of skilled operators is apparent.

So far, the contributions have looked at the application of various aspects of IT to specific manufacturing processes, mainly in the engineering industry. Peter Murray and James Wickham look at the manufacture of IT itself in Ireland. They show what a low technology process the manufacture of IT is, at least in the assembly stages. They go on to argue that despite this low technology production process, the demand for a skilled workforce remains. They argue that this is not because of the intrinsic nature of the process, but a means of selecting 'responsible' workers. So far as the technicians are concerned, where skills are required in the jobs, they show the ways in which management is trying to reduce its dependence on technical expertise through narrowing the job. The implication of this argument is that claims which equate high technology with high skills should be treated with caution.

A number of the contributions have touched on the issue of training. Sheila Rothwell and David Davidson report a number of case studies of the training programmes associated with the implementation of information technology. They found a good deal of variation in the attention given to training, and the amount of training given to the workforce. Even where management were committed to systematic training programmes, they found that such provision was the first to go if the overall programme fell behind schedule. Their conclusion is that not enough attention is given to training, and that the effective utilisation of information technology is thereby undermined.

The contributions so far to this book have examined various facets of the activities of management in relation to IT. Cynthia Cockburn turns to the implications for the worker and trade union organisation of new technology in the newspaper industry, where the press owners are using computerised photocomposition to undermine craft control and the closed shop. While the craftsmen are strong enough in the newspaper industry to hold onto the new-style jobs, elsewhere in print the technology is being used, as it is clearly designed to be used, to enable employers to substitute cheaper and less industrially-experienced female labour. She points out that for women, the new technology is an opportunity to enter some of the jobs in print from which male trade unions have long organised to exclude them. The union is being forced to rethink the concept of skill and its membership strategy.

Five main themes run through these contributions. Firstly, the 'impact' of information technology is contingent on the existing context. Factors such as batch and plant size, the product market environment, national traditions, managerial attitudes, the existing technology, and trade union organisation are all major influences in the way information technology is implemented. Secondly, there is much variety in the forms of organisation associated with the same innovation, and much room for choice in that organisation. The precise form implemented will be the outcome of bargaining processes within the organisation. Thirdly, skills and training are key issues, but there is no single tendency towards deskilling or reskilling. Trade union organisation, management's aims for the new technology, and its precise function all affect the direction of change. Fourthly, the general competence of British management to handle the technical and organisational issues raised by adopting IT for production processes is in doubt. Fifthly, the interests of the workforce have hardly entered into the strategies for technological change adopted by management, and, with the exception of print, workers have been unable to influence technological change positively through their trade unions.

In his conclusions, Craig Littler takes up these points in more depth, and places the present wave of technological innovation in its historical context. Earlier in this introduction, it was argued that the present IT debate is very much a re-run of the automation debate of the fifties. Technological change has been the norm in capitalist economies for the last two hundred years, and for the whole of this century the pace of change has, overall, been increasing. It is crucial for the understanding of the implications of IT that the present period is seen in the light of this history.

NEW TECHNOLOGIES; OLD PROBLEMS

De-industrialisation is the failure of an economy's manufacturing base to provide the wealth to support the desired level of imports in a socially acceptable way(25); the British economy is clearly de-industrialising in this sense. Although it is measured at the aggregate level of the national economy, de-industrialisation is the consequence of actions and inactions at the level of individual manufacturing units having a negative cumulative effect. Case study work, therefore, is crucial to understanding the de-industrialisation process. The themes developed in this book have a number of implications for policies aimed at reversing de-industrialisation.

In the United States, the discussion about de-industrialisation has taken the form of the 'productivity crisis' debate - see almost any issue of the Harvard Business Review. In an important contribution to this debate, Hayes and Abernathy(26) have argued that the reason for this productivity crisis is that corporate decision-making has become orientated towards financial rather than productive goals, and this is reflected in the declining proportion of technologists amongst senior managers. The de-industrialisation debate in this country has tended to be focussed on strategy options at the level of the state, and in particular, the role of the state in promoting technological change. But as John Bessant argues, innovation is an evolutionary process, depending upon the cumulative effects of decisions by relatively junior managers. The implication of this is that the technical skills of the existing management body are a crucial factor in the diffusion of information technology. Both Erik Arnold and James Fleck point to the lack of expertise amongst management, and the Anglo-German team suggest ways in which the British managerial culture may exclude alternative ways of utilising CNC. David Buchanan shows the ways in which the aims of different levels of management may be in contradiction and undermine effective utilisation of IT(27).

Thus the cases presented here support the contention that one of the contributory factors to the de-industrialisation is the low level of technical skills available. This has been strongly argued by Prais(28); indeed he goes so far as to argue that lack of technical expertise is one of the two key factors in Britain's relatively poor performance(29). The evidence of Sheila Rothwell and David Davidson is that management are tending to give training a low priority, and are content to buy in the most important skills. Peter Murray and James Wickham, though, point to an important, but often overlooked, aspect of this issue. It has been argued that when employers complain of skill shortages, they are not always concerned with a lack of technical expertise, but of 'acceptable' workers (30). The Irish team provide evidence that, so far as the assembly workers in the electronics plant were concerned, their qualifications did not provide them with expertise, but proved their acceptability to management.

The competitive pressures that simulate innovation do not, however, determine either the type, or mode of utilisation of the new technology(31). Child has argued that organisations possess a degree of strategic choice - managerial decisions are only bounded by contingencies, not determined, and that the final decision will be dependent upon the power relations within the particular organisation(32). While Graham Winch shows the ways in which contingencies encourage particular organisation designs, David Buchanan emphasises the range of strategic choices that are available.

The most pervasive power relationship within an organisation is that between management and the workforce. Writers in the 'labour process' school(33) have gone so far as to argue that the strategic choices about technology are made with the overriding aim that, in Ure's words, 'the refractory hand of labour will...be taught docility'(34). While there are many contingencies other than labour unrest which restrict management's strategic choices, for instance the product market, the evidence from James Fleck, Graham Winch, and David Buchanan is that such control is an important motive in strategic choice. On the other hand, Cynthia Cockburn shows the way in which trade union organisation, based on a policy of only organising key groups of workers, and leaving out others, in particular women, can greatly influence strategic choices over work organisation.

Strategic choice does not only lie in the way in which the technology is utilised, but also in the choice of technology itself. Rosenbrock has distinguished between a 'tool' which enhances workers' skills, and a 'machine' which replaces and destroys skills. He argues for the development of IT as a tool rather than a machine(35). This argument has been made explicitly with reference to CAD(36), robotics(37), and NC(38). However, Graham Winch's analysis of the integration associated with CAD/CAM suggests that the possibilities of implementing tools rather than machines are being constrained by overall developments. While the arguments for tools are well established for stand-alone innovations, the development of Integrated Manufacturing Systems(39) will tend to centralise control in an hierarchical manner. Strategic choices at the most senior management level are closing off technological options lower down(40).

At the policy level, the trade union response to information technology has been positive(41), and none of the cases reported what could be construed as worker resistance to the principle of changing technology. This is typical of the situation nationally(42). However, the main policy instrument developed by the TUC to handle technological change - the New Technology Agreement(43) - has been notable for its absence amongst workers concerned with manufacturing processes(44). James Fleck reports that management have been cautious where they feared industrial relations problems, and Erik Arnold gives details of the payments for change gained by drawing staff, but in general, as Sheila Rothwell and David Davidson argue, workers have been more concerned with bargaining over the effects of the recession than over aspects of new technology. Only in the case of print described by Cynthia Cockburn have management been obliged to make significant concessions to trade union bargaining power.

Thus trade unions are failing to exercise power in the strategic choices over technological change made by management. This is largely due to the effects of the recession, but it is also symptomatic of weaknesses in trade union organisation, and in particular, of the poor links between blue and white collar trade unionists. These failings mean that information technology is being implemented without proper consideration of the interests of the workers who will use it.

Over 100 years ago, Marx analysed the ways in which the mechanisation of Victorian industry was yielding significant gains for the employers while its potential for relieving the lot of the workers was denied. He pointed to the 'economic paradox that the most powerful instrument for reducing labour-time suffers a dialectical inversion and becomes the most unfailing means for turning the whole lifetime of the worker and his family into labour-time at capital's disposal'(45). The dilemma posed by information technology in the eighties is how to realise the productivity gains that are essential for the reversal of de-industrialisation, while ensuring that IT is used to create tools, and not machines.

REFERENCES

1. BRIGHT, J.R. Automation and Management. Boston: Harvard University Press, 1958, p.4. Of course the etymon is much older. The OED cites the word 'automatical' - to mean automatic - as first being used in 1586, and 'automatic' was in established use in the nineteenth century. The concept of the 'automatic factory' was first sketched out by an engineer working for Morris in the early twenties. See WOOLLARD, F.G. Principles of Mass and Flow Production. London: Iliffe, 1954.

2. BRIGHT, op cit, Appendix 1.

3. This is not the place to attempt a definition of automation, and its distinction, if any, from mechanisation, but any definition should include the notion of the removal of humans from the control cycle of machines.

4. Cited BRIGHT, op cit, Appendix 2.

5. MACMILLAN, R.H. Automation; Friend or Foe? London: Cambridge University Press, 1956.

6. WIENER, N. Cybernetics. New York: The Technology Press, 1948, p.37. See also his The Human Use of Human Beings. New York: Houghton Mifflin, 1950.

7. DIEBOLD, J. Automation: The Advent of the Automatic Factory. New York: Van Nostrand, 1952. Diebold was also an early (1957) user of the term 'information technology' in its present sense. See his Beyond Automation. New York: McGraw-Hill, 1964, Chapters 3 and 4, especially p.63.

8. Cited BRIGHT, op cit, Appendix 1.

9. See, for instance, those cases reviewed in MEISSNER, M. Technology and the Worker. San Francisco: Chandler Publishing, 1969.

10. See for instance, MALLET, S. La Nouvelle Classe Ouvriere. Paris: Editions de Seuil, 1963; and BLAUNER, R. Alienation and Freedom. Chicago: University of Chicago Press, 1964.

11. BRIGHT, op cit, and BRIGHT, J.R. Does Automation Raise Skill Requirements? in: Harvard Business Review. July/August 1958.

12. BRIGHT, op cit, Chapter 3.

13. NOYCE, R.N. Microelectronics. in: Scientific American. 237, 1977.

14. NORA, S. and MINC, A. The Computerization of Society. Cambridge: MIT Press, 1980.

15. COUNCIL FOR SCIENCE AND SOCIETY. New Technology; Society, Employment and Skill. London:CSS, 1981; Clayre, A. (ed) Nature and Industrialization. London: Oxford University Press, 1977, is a fascinating anthology of this debate.

15. NORTHCOTT, J., ROGERS, P. and ZEILINGER, A. Microelectronics in Industry: Survey Statistics. London: Policy Studies Institute, 1982. See also the discussion of these figures in Northcott, J. and Rogers, P. Microelectronics in Industry: What's Happening in Britain. London: Policy Studies Institute, 1982.

16. GIBBS, D., EDWARDS, T. and THWAITES, A. The Diffusion of New Technology and the Northern Region. in: Northern Economic Review. No. 5, 1982.

18. British Robot Association figures.

19. JOLLY, B.S. and GARDNER, A. Comparative Adoption of Automation and Microelectronics in the UK, Germany and Sweden. in: Automation. March, 1981.

20. METALWORKING PRODUCTION. Fifth Survey of Machine Tools and Production Equipment in Great Britain. London: Morgan-Grampion, 1983.

21. GEBHARDT, A. and HATZOLD, O. Numerically Controlled Machine Tools. in: Nabseth, L. and Ray, G. The Diffusion of New Industrial Processes. London: Cambridge University Press, 1974.

22. SORGE, A., WARNER, M., HARTMANN, G. and NICHOLAS, I. Microelectronics and Manpower in Manufacturing. Berlin: International Institute of Management, 1981.

23. SARGENT, R. Process Control and the Impact of Microelectronics. Chemistry and Industry. 20th December, 1980.

24. FRANCIS, A., SNELL, M., WILLMAN, P. and WINCH, G.M. Management, Industrial Relations and New Technology for the BL Metro. Mimeo, Imperial College, 1982.

25. SINGH, A. UK Industry and the World Economy: A Case of De-industrialisation? Cambridge Journal of Economics. 1, 1977.

26. HAYES, R. and ABERNATHY, W. Managing Our Way to Economic Decline. Harvard Business Review. July/August 1980.

27. See also SWORDS-ISHERWOOD, N. and SENKER, P. Management Resistance to the New Technology. in: Forester, T. (ed) The Microelectronics Revolution. Oxford: Blackwell, 1980.

28. PRAIS, S.J. Vocational Qualification of the Labour Force in Britain and West Germany. National Institute Economic Review. No. 98, 1981.

29. PRAIS, S.J. Productivity and Industrial Structure. Cambridge: Cambridge University Press, 1981, p.272.

30. OLIVER, J.M. and TURTON, J.R. Is There a Shortage of Skilled Labour? British Journal of Industrial Relations. 20, 1982.

31. For a general critique of the tendency to deduce organisation imperatives from economic ones, see TOMLINSON, J. The Unequal Struggle? London, Methuen, 1982.

32. CHILD, J. Organizational Structure, Environment and Performance: The Role of Strategic Choice. in: Sociology. 6, 1972.

33. See WOOD, S. The Degradation of Work? London: Hutchinson, 1982, Chapter1.

34. Cited, MARX, K. Capital, Vol. 1. Harmondsworth: Penguin, 1976.

35. ROSENBROCK, H.H. Technology Policies and Options. in: Bjorn-Andersen, N., Earl, M., Holst, O. and Mumford, E. Information Society; for Richer, for Poorer. Amsterdam: North Holland, 1982.

36. ROSENBROCK, H.H. The Future of Control. in: Automatica. 13, 1977.

37. THRING, M.W. Robotics and Telechirics. in: Industrial and Commercial Training. 12, 1980.

38. BOONE, J.E., SATINE, L., HINDUJA, F. and VALE, G. Back to Operator Control? in: Numerical Engineering. 1, 1980.

39. See HALEVI, G. The Role of Computers in Manufacturing Processes. New York: John Wiley, 1980.

40. This argument applies equally to user-oriented flexible manufacturing systems because they will also be dependent upon the output of CAD/CAM systems.

41. See ROBINS, K. and WEBSTER, F. New Technology: A Survey of Trade Union Response in Britain. in: Industrial Relations Journal. 13, 1982; See MANWARING, T. The Trade Union Response to New Technology. in: Industrial Relations Journal. 12, 1981 for a critical review.

42. See NORTHCOTT, J. and ROGERS, P. Microelectronics in Industry. London: Policy Studies Institute, 1982, p. 55.

43. TRADES UNION CONGRESS. Employment and Technology. London: TUC, 1979.

44. INCOMES DATA SERVICES. Changing Technology. Study No. 220, London: IDS, 1980. WILLIAMS, R. and MOSELEY, R. The Trade Union Response to Information Technology. in: BJORN-ANDERSEN, op cit. Labour Research Department. LRD Survey of New Technology. London: LRD, 1982.

45. MARX, op cit. p. 532.

Management and manufacturing innovation: the case of information technology

John Bessant

INTRODUCTION

Concern about the rate of adoption of information technology by manufacturing industry has led the UK government to commit over £1 billion in various schemes aimed at accelerating development and diffusion of this technology(1). Despite this strong pressure, backed by considerable exhortation in the media and especially during IT82, the level of take-up in the UK remains relatively low in comparison to competitors like Germany, Scandinavia, and especially Japan. Examples exist indicating that this pattern extends across the range of applications in manufacturing - robotics, flexible manufacturing systems, computer-aided design, programmable logic control, automated test equipment, automated warehousing and so on(2).

Analysis of the adoption pattern of microprocessor applications - the first of the constituent technologies in IT to receive promotional treatment - indicates that there is wide sectoral variation with (not surprisingly) advanced industries and those close to the technology like electronics showing much higher rates of adoption than more traditional sectors like textiles and clothing(3). This pattern is a source of concern given that it is precisely these traditional industries which are most threatened by their own lack of international competitiveness - and thus most urgently in need of the productivity improving contribution offered by IT.

Large firms - again, not surprisingly given earlier experience of the diffusion of innovations(4) - are more likely to adopt than smaller ones. Though the benefits to the latter if they do take the risk of entry into the technology can be proportionately much greater. More serious, perhaps, is the suggestion that foreign owned subsidiaries in the UK are more likely to adopt than similar British-owned firms - raising again the spectre of the lack of technical progressiveness amongst UK managers(5).

Diffusion theory suggests that the rate of take-up of a new technology will depend on a combination of factors which include its relative profitability, the extent to which the potential user is committed to the present generation of plant, and on various learning effects and organisational variables(6,7). In times of recession we would expect rates of investment to fall with declining profitability - and there is no doubt that this is the single most important reason for the slow rate of investment in IT in the UK. However it does not satisfactorily explain why some firms have been able to innovate in this direction whilst others in similar circumstances - same sector, size, market position and so on - have not. The answer is not to be found in the capital commitment argument either: whilst some large capital intensive industries may be constrained by the present vintage of their plant, the nature of IT is such that it permits relatively low cost options - for example, retro-fitting of controls to existing plant.

It is suggested that, whilst there are undoubtedly many economic and technical issues to be resolved at a national level, it is at the level of the individual firm that the rate and extent of diffusion is decided. For this reason, it is important to look more closely at the influential factors operating within the firm.

This paper reports on a variety of case studies carried out over a four year period for agencies such as the Department of Industry, Anglo-German Foundation, Policy Studies Institute, and the United Nations. It presents some findings on the key diffusion influences within firms concentrating particularly on organisational and management issues. Some comments about the pattern of choices which emerge during the innovation decision process and their implications for diffusion and successful implementation are also offered.

MOTIVES FOR INNOVATION

The reasons which firms advance for having invested in new technology vary widely: most of this investment involves a pattern of incremental process change. This can be classed as 'manufacturing innovation', which can be defined as the way in which firms make improvements to their production process without altering the basic products or manufacturing routes involved(8). Most day-to-day innovation falls into this category and the cumulative effect should not be underestimated: studies of this type of 'bread and butter' innovation have shown the significant contribution to performance which can be made(9).Because it is a regular and to some extent planned process linked to annual investment budgets, it is often the responsibility of operational managers and supervisors - that is, much closer to the production process itself rather than at board level. In this respect it involves much more tactical decision-making, as opposed to the strategic type associated with major process innovation. This distinction is significant because it means that the adoption and implementation of this type of innovation will be more strongly influenced by local level issues - fit with existing pattern of production, local tradition of innovation, and the previous success or failure of innovation, industrial relations, workforce attitudes to technical change and so on.

Many IT-based innovations belong to this group: despite the novelty and technological sophistication often associated with such equipment, there is little basic change in the production process involved. For example, using a robot in the car industry to weld bodies or spray paint may be a novel and impressive technological change, but the basic process of making cars by welding metal and painting it remains the same. From the point of view of the engineers and designers involved, IT often represents little more than a sophisticated new set of tools for carrying out the same tasks. This perspective on IT as an evolutionary rather than revolutionary change differs markedly from the presentation of the technology in the media, and it is important in its implications for the adoption process.

The economist's view of technical change argues that some form of relative advantage or profitability will be needed in order to provide the incentive for adoption. In practice the range of motives firms put forward for this relative advantage is wide: Table 1 gives a list of the more common ones. Significantly very few firms indicated that they had a single clearly defined motive for moving into the technology: in most cases it was a combination of several factors which was used in the investment justification.

This fits with another aspect of manufacturing innovation - it is a regular process of continuing improvement. Irrespective of the industries in which they work, engineers tend to have 'ideal' models of how production ought to be operating, against which they assess the needs of the actual operations. Such models have components which include continuity, flexibility, minimal factor costs/inputs and maximal output, high quality, high levels of control over all process stages and so on. In reality it is only the advanced process industries like petrochemicals which come anywhere near this ideal but it can be argued that the presence of an internal model of 'optimal production' is a strong influence on the direction and choices about technical change within the firm.

The pattern of manufacturing innovation within the firm will reflect strongly the progress towards achieving the ideal characteristics of the optimal model for a given process. That is, the organisation will constantly undergo a kind of 'tuning' of production activities, much as vehicle engines might be tuned to optimise performance. Priorities will vary over time as a result of changes within and without the firm but in general, engineers will seek those technological options which bring them closest to optimal production. They will also tend to favour innovations which contribute in several different ways - as can often be the case with IT-based applications. For example, computer-numerically controlled (CNC) machine tools under certain circumstances can be both labour and capital saving, can increase flexibility and machine utilisation, can offer better quality and consistency, work to finer tolerances and are more reliable and easy to maintain than conventional tools.

Table 1: Common motives for adoption of IT based manufacturing innovations

- savings on direct labour costs
- savings on skilled labour (coping with skill shortages by using the capacity of IT to embody skills within the software)
- savings on indirect labour (through improved reliability, easier maintenance, online monitoring, etc)
- improvements in machine operation - greater accuracy, flexibility, etc
- reduced cycle times
- shorter set-up times
- improved reliability, easier maintenance
- improved production control, better information availability
- energy savings
- material savings
- space savings
- improvements in process safety

As progress in manufacturing innovation brings the firm closer to the optimal model so the costs (in research and development, manpower and so on) rise sharply, following a pattern of diminishing returns. This has been noted by economists such as Soete(10) and is typical of the state of the large capital intensive process industries where considerable investment goes to getting small percentage improvements in performance. An example of this is the use of a microprocessor controller to achieve even a half-percent energy saving through tighter control: this investment can be justified in terms of the very large throughput volumes. In a more general sense, though, such innovations must be seen as very fine tuning: as the 'technological slack' is taken up and the costs rise, so alternative processes become attractive. When a new process appears the cycle begins again with a new phase of manufacturing innovation: analysis of the technological process development of most major industries supports this cyclic model. It has a value for us in looking at IT-based technologies because it suggests where and what innovations might be involved and what the likelihood of adoption by engineers is going to be. Thus advanced industries which are quite a long way along the manufacturing innovation curve and already have developed traditions of sophisticated controls will tend to be early adopters but the impact of the technology will not be dramatic - either in terms of process economics and technology or of workplace and organisational resistance. By contrast, industries which are at an early stage on the curve - for example, batch engineering - stand to gain much more from using IT but it will involve considerable adaptation and change on the part of the adopting organisation.

It is interesting to speculate on the source of the internal model which engineers and production staff use: it must become developed during the period of formal education and as a result of accumulated experience and accepted practice. Analysis of the education systems of other countries - for example Japan or Germany - shows that there are major differences with these countries providing a much more broadly-based education which emphasises the systemic nature of production(11). It has been argued that this integrated view of production leads to approaches to production management which minimise the interdepartmental conflicts characteristic of much of UK industry(12). Since IT is essentially an integrating technology, it can be argued that those countries whose engineers and production managers hold internal models of a more systemic kind will be better placed to take advantage of the opportunities offered by this technology.

DIFFUSION PROBLEMS AT THE LEVEL OF THE FIRM

It is clear from the case studies that slow diffusion cannot be accounted for by a single causative factor or group of factors: as previous work on diffusion of process innovations has shown, it is a complex pattern(13). On a qualitative basis we can isolate some of the more general factors, which include the following:

- investment pressures: under present economic conditions, the level of profitability of most firms in manufacturing is extremely low and this has clearly had a major effect on the rate and level of investments which are made.

- lack of suitable technology: whilst IT has developed rapidly there are still a number of areas where application is slowed by the absence of suitable technology, in particular transducers (sensors and actuators to relate the production process to the IT system and back again).

- lack of applicability: despite the high degree of flexibility available through the programmable nature of IT, there are some areas where its application is limited on grounds of technical incompatibility, lack of economic justification or that the task involved is too simple and to use IT would be 'using a sledgehammer to crack a nut'.

- vintage of present capital: for large, capital intensive firms the investment replacement cycle may be quite long and it may be that IT whilst offering attractive advantages cannot be economically introduced until the next phase of replacement.

- firm size: it is clear that large firms are able to devote more resources - whether financial, technical or personnel - to innovation and in a small number of cases the lack of available resources constituted a barrier to innovation using IT for smaller firms.

Leaving these factors aside, we are left with a group of organisation-level factors which have a strong behavioural component. These can be broadly grouped under the following headings which are essentially related in some way or other to compatibility - that is, to getting a good fit between the technology and the organisation. We will discuss each briefly:

- awareness and information
- training skills and manpower planning
- industrial relations
- attitudes to new technology

Awareness and information

In 1979 a MORI poll(14) of senior managers revealed that very few were aware of microelectronics technology or planning to use it in their products or processes. After several years of considerable exposure and a strong awareness and education programme under the auspices of various government schemes (culminating in IT82), the results of such polls are more encouraging. It is now clear that awareness of the technology is good, but this has opened up a new problem associated with applicability. The difficulty is that whilst it is easy to appreciate that IT is a powerful and interesting technology but quite another thing to spot ways in which it can be used in a particular firm to advantage. As a recent report put it, '...potential users have, as yet, insufficient knowledge and understanding of the specific applications of the new technologies in their own sectors...'(15). This raises some important points. First there is a need for a developing education programme aimed at applications skills: an example of this is a recent venture by the National Computing Centre(16). Beyond this there is a need for specialist training - not only to develop technical skills associated with the new technology, but also to improve understanding of the characteristics (and particularly the controllable mechanisms) of existing manufacturing processes within various sectors. It may well be that the interdisciplinary skills needed at this level will take some time to develop and in the short-term firms should consider making use of expert consultants to provide this. Many of the government schemes, like the Microprocessor Application Project, lay considerable stress on this early phase of awareness and training: the availability of some financial support for exploration and feasibility study may help to remove barriers to adoption.

Training, skills and manpower planning

Clearly any move to new technology will involve some changes in the skill profile of the company. New and modified skills will be required - from operators faced with new procedures and equipment, through to maintenance staff who will have to cope with sophisticated electronics and diagnostic systems, to management, faced with the responsibility for overseeing the change. There will be very few organisations which possess the right skills mix at the outset: consequently there are major implications for manpower and training policies.

One option is to recruit, but this may be a problem because there are severe shortages of many of the required skills - such as electronics engineers, instrument mechanics and other technician grades. Even if these skills can be recruited they may well be expensive because of the demand in the market. More serious is the fact that such staff will be unfamiliar with the company's operations and environment: absorbing these to the point where new staff can make a useful contribution can be costly in terms of time.

An alternative is to sub-contract those specialist tasks for which skills are in short supply - for example, maintenance work. Again, this is likely to be expensive and there is still the lack of knowledge about a particular company's operations to be considered. Additionally, the degree of control in a field like maintenance is considerably less than if the company had an in-house capability - something many firms are reluctant to accept.

Another option is to train staff within the company to meet the new skills required. There are still problems of cost, of time spent on training rather than productive work, of training resource provisions, and so on. However the big advantages lie in the familiarity such staff have with the company - once trained they can be useful quickly - and the control the company retains by having the necessary skills in-house. Using this option also means that relocation of staff displaced by new technology from other areas is possible: this helps develop a flexible, high quality workforce and cushions the impact of change on the organisation.

It is a matter of choice which way the firm tackles the problem of providing skilled resources to support new technology. However it is clear from the above that one of the most important guidelines is 'anticipation': whichever route the company uses, it must make its provisions ahead of the investment in equipment. A related issue concerns the training function which is often given a low priority, especially in certain manufacturing sectors. While this might have been something firms could get away with in the past it poses a number of problems (and misses out on a number of advantages) for the future.

First, the trend in manufacturing has been one of increasing capital intensity - a smaller number of people are becoming responsible for an increasingly large amount of plant and equipment of high sophistication. This is already the pattern in the large scale process industries where in oil refineries, for example, entire operations are left to the discretionary control of a small group working from a central control room and relying on high levels of automation for the bulk of plant operation. With the emergence of information technology, new possibilities have been opened up in the field of manufacturing technology and we are seeing a move towards the process industry pattern across a wide range of industrial sectors, and this places greater emphasis on the need to invest in training.

Second is the fact that the trend is also towards greater integration of production operations. For example, in metalworking, the shift has been towards complex machining centres which can handle a variety of operations within a single machine under programmable control. Systems also exist to link - via robotics or programmable conveyors - different machines within a production cell. Further up the line we see the stockholding of parts and finished products being tied in, again under computer management. Another level up brings in the production and stock control parts of the management information system - and so on. There is also increasing integration between different areas in manufacturing - for example, the linking up of design and manufacture in CAD/CAM (computer-aided design and manufacturing) systems. Such a shift in the pattern of manufacturing - which Halevi(17) has called 'HAL-automation', from the Hebrew word meaning 'all-embracing' - has considerable implications for skill structures within the firm.

Evidence from the case studies suggests that the requirement is for several different types of skill, some new and others representing new combinations of existing skills. For example, the maintenance of industrial robots requires a mixture of electronics, electrical, hydraulics and mechanical skills as well as a new approach to diagnosis and fault repair(18). Similarly, machine tool operators become responsible for programming and editing of the control programs for their machines as well as operating them. Whilst it is not clear exactly what new or different skills will be involved it is clear that more training and updating of training is going to be needed(19).

With changes of the kind described above it may be necessary to revise approaches to the process of training itself. Traditionally training takes place at the end of the innovation cycle: the hardware and software have already been developed and commissioning is imminent. Training inputs are often 'plugged-in' at this late stage - which means that much of the early operating life of the plant is spent of costly on-the-job learning before the operators and support staff know enough about it to run it well. This type of training may be easy to plan - especially if the inputs are provided by the equipment supplier as part of the package - but may not be the most appropriate for the actual needs of the new plant. An alternative is becoming increasingly popular and that is to allow for training to be run in parallel with the innovation process itself. This offers a number of advantages: first and most important, it allows for much more flexibility within the training process. Given that many firms do not know in detail what skills they need to support new technology, what the mix between breadth and depth should be, how many multi-skilled staff will be needed and so on, it follows that any process which allows for changes and tailors the training to the evolving needs of the project will be an attractive option(20).

Second there is now a growing body of evidence to show that 'participative' design - that is, involving the users in some way - produces a better and more workable system(21). Changes which would be expensive to rectify in the post-commissioning phases can be designed out much earlier. For example, the users have a great deal of experience of the actual operating conditions and may be able to suggest modifications which cope with unforeseen problems. By building training into this process, the need for supplementary training to fill gaps and weak spots (which might get missed out of traditional plug-in types of training) can be eliminated. More important, the involvement at the design stage of those who are going to work the system means that they will be familiar with it and able to deal with many of the out of the ordinary operating situations, using experience gained during the development process. Furthermore some of the problems of resistance to the introduction of new technology can be avoided: instead the workforce feel a sense of 'ownership' - it becomes 'their' plant, and they will be motivated to make it work.

Industrial relations

Given the potential impact of IT on the quantity and quality of work, it might be expected that trade union hostility would represent a major barrier to diffusion. However, the actual trade union response has been strongly positive: in policy statements(22) the reasons for this become clear - the threat to jobs from declining international competitiveness is greater than that from adoption of new technology. It may also be possible to exert some control over the way in which new technology affects the workplace. The main approach is to try and achieve greater involvement in the decision processes surrounding the introduction of information technology: one mechanism advanced for this has been the New Technology Agreement (NTA).

The bulk of NTAs so far signed have been with white collar unions and in the advanced sectors of manufacturing like electronics and aerospace(23). More traditional sectors have stuck by established collective bargaining practices although there is some concern that the effectiveness of these in the future may be limited - for example they might be unable to cope with problems raised by greater integration of tasks, crossing of demarcation lines and so on. In the case studies industrial relations was not considered to be an important barrier to introduction of IT although a few large firms expressed some fears of union hostility and said that this anxiety had led them to hold off on investing in IT in certain 'sensitive' areas. The general view was that consultation and information provision about technical change were part and parcel of good industrial relations practice and most firms paid attention to this. The same went for introduction strategies: the bulk of firms had used some form of 'selling' of the technology to the workforce, although there were few examples of actual participation in the design process. It should also be borne in mind that the lack of resistance to change on the part of trade unions may be related to their relatively weak position in manufacturing as a consequence of recession, where fears of unemployment have had a marked effect on attitudes.

Attitudes to new technology

Inevitably there will be a degree of resistance to new technology: this forms part of a more basic fear of change which can be observed in any population. Information technology, however, is particularly threatening in this connection since it is such a novel technology, invisible in operation and has appeared rapidly - over less than ten years as far as most users are concerned. Within the firm it is important to appreciate that it is the perceived rather than the actual characteristics of the technology which influence attitude formation. That is, if a person thinks the job will change for the worse as a result of the adoption of IT, then that person will be hostile to its introduction - irrespective of the facts of the case. There are a number of characteristics which are important in this connection, including:

- complexity: there is little doubt that IT is seen as a complex technology which remains very much the province of specialists. Even amongst the relevant specialist disciplines there are barriers between different knowledge areas, and for more 'conventional' technologists, the level of understanding is confined to broad outlines only. At the level of operations and support staff and in management there is little understanding about how the technology works: this picture is further complicated by the highly specialised jargon associated with the technology. The effect of this complexity can be to predispose potential users against the technology because it appears to lie beyond the firm's capacity to understand and use it. This is in line with innovation theory where subjective judgements of this kind can have a disproportionate effect: in some cases studied it was clear non-adoption represented what might be termed as a 'not applicable here' approach analogous to the 'not invented here' phenomenon.

- observability: closely related to complexity in predisposing negative attitudes is the 'black box' nature of IT. In operation the technology is completely invisible, with no apparently moving parts, no indication of status, whether and where there is a fault, and so on. The way in which control is achieved remains invisible to the operator. Responses in some of the case studies indicate the concern felt: in one firm microprocessor-based sequence controls were rejected in favour of pneumatics - despite their apparent superiority for that application. The production manager justified his decision by explaining that pneumatics were easy to understand, familiar and simple to repair, that it was easy to follow the control circuitry and that in the end it was matching the control system to the operators which mattered, and not vice versa.

- novelty: another problem relates to the speed at which IT has arrived: for most employees changes in control, technology have come about with increasing rapidity over the past years, and it appears that their rate of adaptation is out of step with these. As an example, a middle-aged maintenance craftesman may well have begun his career working on systems based on valve logic: since then he will have seen changes to relays, to transistors, to hard-wired integrated electronics, and now to programmable devices of growing complexity. This age/novelty problem emerged frequently in the case studies and led to many firms indicating a preference for younger employees for training to work on IT based systems.

- lack of technical confidence: in some cases concern was expressed about the technical capabilities of IT and particularly their reliability. In one case a group of production managers were hostile to a proposed process control application: they felt that familiar operating procedures had always produced acceptable results and that previous attempts at technical change had not always been successful. Their response was to press for considerable overdesign of the system, with many back-up systems and a high level of redundancy built in to guarantee reliability, inevitably raising the cost of the project.

CHANGES IN WORKING PATTERNS AND STRUCTURES

The introduction of IT is likely to bring about a number of changes in the way the firm operates: the extent of these depends on the scale of innovation involved. Clearly, buying a small microprocessor-based laboratory instrument will have minimal effect - whereas a major investment in converting a batch plant to continuous operation under direct computer control will have implications for the whole organisation.

At the operational level, the problem is mainly one of skills and work organisation: one area of particular concern is the location and change in the type of tasks involved. In the process industries, for example, the role of the plant operator has changed radically from a direct involvement with a particular item of plant to an overall 'policing' role in front of a display panel which may be located some distance from the actual plant. It has been suggested that there are negative effects associated with this, not only in terms of ergonomics (for instance, response times under stress to danger signals) but also in the loss of the operator's intuitive understanding of the plant and process being operated.

Overall, such operations-level changes are going to require anticipative job design to avoid problems in implementation. A similar picture emerges for other activities: for example, maintenance patterns change considerably as a result of shifts in skill requirements and the rising level of integration in production so that the distinction between production and maintenance work becomes blurred. Changes include the move to multi-skilled work - as in the maintenance of robots - and the use of internal diagnostics in the system to simplify first line fault-finding. This latter course also permits the firm to operate a 'replace rather than repair' policy for its immediate needs interchanging parts and circuit boards and sending them away for repair under contract.

Similarly supervisory tasks will be affected by the growing use of on-line monitoring techniques. Traditionally supervisory work involves collecting, analysing and transmitting information about activities on the shop floor and then taking discretionary action to keep production moving. Additionally such a position of information control carries with it considerable influence and power within the organisation. Consequently there is likely to be concern about these changes from supervisory staff, and in at least one case there was industrial action blocking the introduction of such an on-line system because of its anticipated effects.

Management changes are more difficult to establish although there is some evidence that decentralisation of operational decision-making has taken place, especially as a result of using on-line monitoring systems. More managers are able to access data through the proliferation of display terminals in offices and on the shop floor, and this again may affect traditional roles which involved one person of group controlling information flow and availability. It seems likely that the future pattern of highly integrated production - with technologies like flexible manufacturing systems as their precursors - will increase the impact on middle management because it brings down to the operational level the management control systems as well as the physical controls.

As far as work organisation is concerned, there is a growing body of evidence to suggest that there is no fixed pattern, predetermined by the technology itself, but rather a range of choices. With CNC technology, for example, these options might extend to the number of shifts worked, the level of skills employed, the location of responsibility for tasks like maintenance and programming, and so on. As Sorge _et al_(24) point out, '...All our results serve to stress the extreme malleability of CNC technology.....the malleability of CNC technology shows in the fact that its technical specification in detail, and its organisational and labour conditions are closely adjusted to company and departmental strategies, existing production engineering and organisation strategies and manpower policies....'

This is supported by recent evidence from West Germany(25) which suggests that in a sample of over 300 firms at least five basic patterns of work organisation were possible. Neither is the pattern confined to CNC technology: research in other sectors reinforces the view that there is considerable choice available in work organisation design(26).

DESIGN SPACE AND STRATEGIC CHOICE

One of the characteristics of the diffusion factors mentioned earlier that they do not represent blocks so much as delays to the adoption of IT based systems. With suitable responses they can be overcome - sooner or later, and more or less easily. They are essentially learning effects when viewed from a macro-level perspective. This matter of degree and rate is critical in technical change because it introduces the concept of choice. Not only are there many different ways of doing things, but it is also impossible to prescribe a single 'best' solution to a given problem because the effects of so many organisational variables - history, technology, tradition of innovation, size, structure and so on - must be taken into account. So we are faced with the problem of trying to find the most appropriate solution for the firm in its unique context - a contingency approach.

Such a theoretical perspective is not new: in many other areas of organisational behaviour it has become accepted as an alternative to theories which do not account for the wide variety of observed factors. Examples include leadership and management styles(27) and organisation structures(28): in connection with this last area of research, Child(29) has developed the concept of 'strategic choice'. Arguably this is a concept which we can apply to our consideration of the options open in the introduction of technical change within the firm.

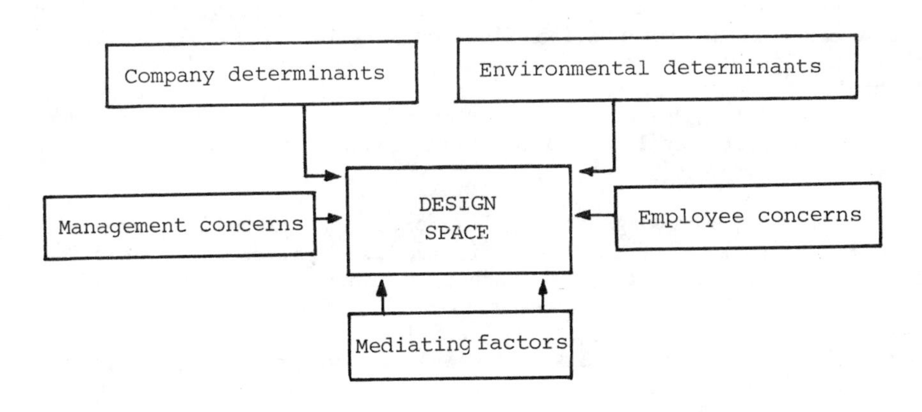

Figure 1 Design Space Model

All of the above factors require some management strategy to deal with them: even the decision to ignore them is a kind of strategy. What we are aiming for is some way of describing the types and extent of choices open to managers and other decision-makers within the firm. In an earlier paper(30) the idea of 'design space' was put forward as a means of describing the room to manouevre in any particular technical change. Figure 1 gives a diagrammatic representation.

In the diagram the space is shown as being shaped by five classes of variables, but this is to some extent an arbitrary classification. The important point is that with any new technology (and particularly with IT) the design space is shaped partly by existing forces and partly by interventions. This is essentially the view taken by Wilkinson(31), who argues that technical change is a negotiated multilateral process involving many actors.

The factors which make up the model are:

- company determinants: these include factors like company size, industrial sector, organisational structure, type of production process (including variables like batch size, flow rates, line balancing requirements and so on). Variables of this kind are subject to management decision and strategic choice but at a higher level and as far as design space is concerned there may be little room to manouevre except in the case of a new greenfield site project. In an existing organisation the main determinant of design space will be that of 'optimal production' pressures to innovate - as discussed earlier.

- environmental determinants: much of what a firm can and cannot do in respect of innovation is influenced by environmental factors which include the overall economic climate, behaviour of competitors, state of the market, availability of resources (materials, energy, labour, etc) and government policies (both to promote technical change - such as subsidies - and to control activities - such as in emission control legislation). Like company determinants, these factors are not usually open to change and must be taken as 'given' for any design space.

- management concerns: in addition to the formal tasks which management is called upon to perform there are also implicit factors which influence decisions about new technology adoption and implementation which include educational and experience background, personal motivation to change, attitudes towards innovation (technical progressiveness) and so on.

- employee concerns: these factors include those which are of interest or concern to the employee in his job and which may be changed as a consequence of innovation. They include job security, relocation and retraining, deskilling, changes in job content, changes in working patterns, changes in informal arrangements, anxiety about aspects of technology (as in the 'Big Brother' side of IT) and so on.

- mediating factors: all of the above factors are largely fixed at the outset of the technical change process (although in the long term they may be altered). However, this last group represent 'dynamic' activities which can directly influence the process of change, and thus the shape of the design space. Included here are factors like the pattern of information flow and availability, training policies, organisational development initiatives, strategy for introduction (participative or autocratic), trade union policies and negotiating power (including NTAs) and so on.

The value of a model like that of design space is that it provides an indication of the complexities involved, and also one explanation for the wide range of responses to new technology which have been observed. It also suggests - since the amount of space available to manouevre in is a variable - what factors might be used to expand or contract it and what scope there might be for improving job or work organisation design processes. It is here that the notion of the 'malleability' of IT becomes important: since it follows that the more a technology lends itself to being shaped, the larger the available design space. This gives support to the view that new technology can be shaped to suit jobs and conditions rather than the other way around. Whether it is in fact used in this fashion is a wider question.

CONCLUSIONS

This paper has indicated some of the organisation level factors associated with the diffusion of IT based technology in manufacturing industry. In particular, it has shown that there a number of areas in which firms exercise strategic choice, and that the pattern of choice has an effect on the rate of diffusion. Whilst it is not possible to prescribe, it does appear that successful implementation is associated with those choices which make use of the flexibility and design space offered by information technology - even if this solution is more costly in terms of time and resources in the short term. Given that the level of investment in manufacturing industry is continuing to rise whilst the numbers employed declines, it seems reasonable to argue for greater attention to be paid to the issue of trying to achieve the best fit between the technology and the organisation.

REFERENCES

1. See, for example, Microelectronics: The New Technology. London: Department of Industry, 1978.

2. JOLLY, B. and GARDNER, E. A Comparison of the Adoption of Automation and Control in the UK, West Germany and Sweden. Aldershot: Systec Consultants, 1980.

3. NORTHCOTT, J. Microelectronics in Industry. London: Policy Studies Institute, 1982.

4. NABSETH, L. and RAY, G. The Diffusion of New Industrial Processes. London: Cambridge University Press, 1974.

5. CARTER, C. and WILLIAMS, B. Industry and Technical Progress. London: Oxford University Press, 1957.

6. MANSFIELD, E. The Economics of Technical Change. New York: Norton, 1968.

7. DAVIES, S. The Diffusion of Process Innovations. Cambridge: Cambridge University Press, 1979.

8. BESSANT, J. Influential Factors in Manufacturing Innovation. Research Policy, 11(2), 1982.

9. HOLLANDER, S. The Sources of Efficiency. Cambridge, Mass.: MIT Press, 1965.

10. SOETE, L. Technical Change, Catching Up and the Productivity Slowdown. in: GRANSTRAND and SIGURDSON, (eds). Technological and Industrial Policy in China and Europe. Lund, Sweden: RPI, 1982.

11. LAWRENCE, P. Managers and Management in West Germany. London: Croom Helm, 1979.

12. NATIONAL ECONOMIC DEVELOPMENT OFFICE. Toolmaking: a Comparison of UK and West German Companies. London: Gauge and Tool Sector Working Party. NEDO, 1982

13. NABSETH and RAY op cit.

14. MARKET AND OPINION RESEARCH INTERNATIONAL. Microelectronics: the views of British Senior Management. London: MORI, 1979.

15. NATIONAL ECONOMIC DEVELOPMENT OFFICE, Industrial Performance: Industrial Applications of Advanced Technologies. Memorandum No. 3(80) for Director. London, NEDO, 1980.

16. NATIONAL COMPUTING CENTRE. Profit from Microchips. Manchester: NCC, 1983.

17. HALEVI, G. The Role of Computers in Production Processes. Chichester: John Wiley and Sons, 1980.

18. FLECK, J. The Introduction of Industrial Robots. London: Frances Pinter, (forthcoming).

19. SWORDS-ISHERWOOD, N. and SENKER, P. (eds). Microelectronics and the Engineering Industry: the Need for Skills. London: Frances Pinter, 1981.

20. VERSTOEP, N. and BESSANT, J. Making It Work: The Role of Training in Manufacturing Innovation. to appear in RIJNSDORP, J. and IMMINCK, S. (eds). Training for Tomorrow. Oxford: Pergamon Press, (forthcoming).

21. MUMFORD, E. Social aspects of systems analysis. Computer Journal, 23(1), 1980.

22. TRADES UNION CONGRESS. Employment and Technology. London, TUC, 1979.

23. WILLIAMS, R. and MOSELEY, R. The Trade Union Response to Information Technology. in: BJORN-ANDERSEN, N., EARL, M., HOLST, O., and MUMFORD, E. (eds). Information Society: For Richer, for Poorer. Amsterdam: North-Holland, 1982.

24. SORGE, A., HARTMANN, G., WARNER, M., and NICHOLAS, I. Mikroelektronik und Arbeit in der Industrie. Stuttgart: Campus Verlag, 1982.

25. REMPP, H. (ed). Wirtschaftliche und Soziale Auswirkungen des CNC-Werkzeugmaschineneinsatzes. Eschborn: RKW, 1982.

26. BESSANT, J. and DICKSON, K. Determinism or design? in: Design '82, Institution of Chemical Engineers Symposium Series, Rugby, 1982.

27. FIEDLER, F. A Theory of Leadership Effectiveness. New York: McGraw-Hill, 1967.

28. SCHREYOGG, G. Umwelt, Technologie und Organisationsstruktur. Stuttgart: Haupt Verlag, 1978.

29. CHILD, J. Organizations. London: Harper and Row, 1977.

30. BESSANT and DICKSON. op cit.

31. WILKINSON, B. The Shopfloor Politics of New Technology. London: Heinemann, 1983.

Information technology as a technological fix: computer aided design in the United Kingdom

Erik Arnold

There is not one determinate relationship between a technology and society. We often speak, however, as if such a relationship existed - for example, we talk about the 'social implications of computers', as if certain consequences inevitably follow from the adoption of computer technology. This is wrong, and if we think about it we know it is wrong: circumstances alter cases. Nonetheless, technological determinism is implicit in analysis across a wide political spectrum. It leads to a dangerous kind of policy prescription: the technological fix, involving the idea that you can simply take a technology from the storehouse of possibilities and successfully apply it (usually without a great deal of further thought) to the problem in hand.

This paper argues that computer aided design (CAD) has been, and often still is, treated in the UK as a technological fix. It is applied as a remedy for some of Britain's industrial ills, yet it is unlikely to be the desired panacea. Indeed, a panacea never does work, precisely because it is resorted to in circumstances where the disease is not well understood. It is not proposed to attempt a comprehensive diagnosis here. Important partial diagnoses have been made by Pavitt(2), Wiener(3) and others. However, it is possible to specify why CAD on its own cannot work, and to set out some elements of a fuller course of treatment.

While computers have many uses in engineering and other forms of design, interactive graphics CAD - where the designer draws with the aid of a computer to define design geometry - has been a focus of government policy and of rapid market growth in recent years. This paper therefore relates to interactive graphics CAD. After some general remarks about the characteristics of successful innovations, US and UK government policies for CAD are compared. It is argued that US policy has been more conducive to the generation and adoption of CAD technology. Some of the difficulties inherent in adopting a new technology are then discussed. Finally, conclusions are drawn about policies for promoting technically progressive industry.

INNOVATION AND COMMERCIAL SUCCESS

At the heart of government intervention in industrial technology is the idea that change, or the pace of change, can be forced. Within economics, the idea that new products can be inflicted on users has strong antecedents. Galbraith has argued that with respect to new industrial products 'the producing arm reaches forward to control its markets, and on beyond to manage the market behaviour and shape the social attitudes of those, ostensibly, that it serves'(4). Schumpeter also said as much: 'It is ... the producer as a rule who initiates economic change, and consumers are educated by him if necessary; they are, as it were, educated by him to want new things'(5).

These analyses assume that invention, innovation and eventual commercial success occur in a linear fashion, without significant influence from users. Other work suggests that inventions are made in response to needs of the demand side, both in capital goods(6) and consumer goods(7). This 'need-pull' position has been supported for innovation(8), but the method and findings have been criticised as inherently biased towards this conclusion(9).

Other studies support more mixed conclusions about demand and supply side factors in successful industrial innovation. In recent British work(10,11) it was clear that _coupling_ technological opportunity to market needs was central. Freeman concludes that 'the crucial contribution of the entrepreneur is to link the novel ideas and the market'(12).

Elsewhere, it has been suggested that _learning_ on the demand side plays a crucial part in the adoption of new technology(13,14). Supplier firms are rarely able to force the pace of this learning, or to ensure that it takes place at all. Governments, on the other hand, often are. The US government, in particular, has encouraged this type of learning through military research and development (R & D) and procurement. Where government desires to promote the development and use of new technology, it will necessarily be incumbent upon it to promote the _coupling_ of supply and demand.

CAD IN THE USA

The first important use of interactive graphics was in the 1950s SAGE early warning radar system in the US. Space and military funding remained major sources of US interactive graphics technology until about 1970. This parallels the story of microelectronics, where the industry was driven by military demand until some time during the later 1960s, when technologies became increasingly orientated to commercial markets. In the 1950s, the Massachusetts Institute of Technology (MIT) produced techniques for numerical control (NC) of machine tools and the APT NC part programming language. This work was funded by the US Air Force (USAF), which had an interest in sophisticated machining techniques for production of complex shapes in aircraft manufacture.

At a meeting in 1959 at MIT, the idea of extending the NC work upstream into computer graphics for design was formulated and the specification of a simple CAD system written (15). The USAF group that paid for the NC research also funded this CAD work. By 1963, a 'man-machine communication system' called SKETCHPAD had been developed(16).

Through the 1960s, the Aerospace Industries Association was a major source of funding for work at MIT in collaboration with the Department of Defense (DoD). By 1965, IBM, McDonnell, and Boeing were experimenting with graphics-based design using refresh screens driven by mainframe computers, and links to NC tools. During the 1969 recession, a good deal of the CAD work in the US aerospace industry was dropped, and this - together with the merger of McDonnell and Douglas in 1967, which left much of the Douglas design team redundant - provided a pool of skilled manpower, some of whom were instrumental in establishing the CAD industry in the 1970s.

The microelectronics industry also needed interactive graphics from the late 1960s, when large scale integration (LSI) made circuits so complex that it became impossible to keep track of designs without computer aid. Some developed CAD internally, while others bought equipment in. Calma, the dominant microelectronics CAD supplier since 1970, entered in response to direct approach from Intel in 1968. The US DoD's current Very High Speed Integrated Circuit programme is likely to reinforce US suppliers' competitive advantage in microelectronics design.

US policy for CAD can be understood in terms of military demand for advanced aerospace products leading to military funding of work on advanced manufacturing technology. The generation of the new technology was *coupled* with the creation of a market for it. In part, this led the military to pay for the development of technology which was then exploited in commercial markets. For example, Lockheed's CAD software (CADAM) was developed under DoD contract but is marketed under licence by IBM. In part, however, it promoted the formation of a pool of technologically knowledgeable manpower in US universities, the aerospace and electronics industries, forming the basis of entrepreneurship and speeding the learning process among potential CAD adopters.

Shimshoni has documented the role of such a pool of skilled and mobile people in diffusing new electronics technology and promoting entrepreneurship in the US scientific instruments industry(17) and, indeed, the mobility of Silicon Valley designers is notorious. The DoD has taken a blunt attitude to the diffusion of technology among potential users by making procurement contracts conditional on NC manufacture, and placing CAD software development contracts with major airframe manufacturers.

TECHNICAL FOUNDATIONS OF A MASS CAD MARKET

The interactive graphics work of the 1960s was done on large mainframe computers. This, combined with the heavy processing demands of the refresh screens used, made CAD prohibitively expensive for most potential industrial applications. Around 1970, four crucial innovations formed the basis of the product recipe adopted by successful CAD manufacturers of the 1970s. These were:

-cheap '2nd generation' minicomputers
-the low-cost (Tektronix) storage tube
-structured programming
-virtual memory.

Together, they allowed massive CAD programs to be developed and crammed into small, cheap computers.

These technologies permitted the packaging of CAD know-how into 'turnkey' systems, opening the possibility of making sales to 'naive' users at a clearly defined all-in price in the hundreds of thousands of dollars. Previously, an investment of above a million dollars was needed for a single CAD terminal, and - since the hardware and software were not packaged together - potential users had difficulty both in identifying the actual cost of CAD and implementing it.

CAD IN THE UK

The UK lagged the US in the development of CAD by several years in the 1960s. There were some 100 interactive graphics terminals in the USA in 1968(18),compared with perhaps 10 in the UK. In 1966, the UK National Engineering Laboratory proposed that a national design centre be established, and in 1967 the Ministry of Technology founded the CAD Centre (CADC) at Cambridge, based around an Atlas 2 computer. Thus, while the US was driving CAD technology via close military-industrial links, the UK set up a government laboratory. Hence, the UK has possessed small numbers of good CAD designers, but not the larger pool of manpower needed to create an industry or to promote the exploitation of the technology in a broad range of industrial sectors.

In the UK at the end of the 1960s, Elliott Automation, Ferranti, ICT and Marconi were offering systems or services using their own interactive graphics hardware, and Racal were offering services based on Elliott Automation hardware. Elliott's graphics terminal was abandoned after the reorganisation of the UK computer industry in 1970. The newly formed ICL (previously ICT) also withdrew, and the Marconi product disappeared along with the Myriad computer. After some success in the mapping sector, Ferranti withdrew from the systems market until 1980. This market was therefore left to US companies. In the event, this was the worst possible time for the UK to pull out, because the technical developments (outlined above) which would enable CAD to become more of a mass-market product were just then taking place.

The contrast with US policy is marked. The CADC provided a centre from which it was hoped expertise and entrepreneurship would diffuse, and indeed some small Cambridge-based companies have been formed by ex-CADC personnel. But there was no mechanism (beyond consultancy) forcing CADC personnel into industry and education or forcing industry to take notice of CAD. Hence, despite the availability of some CAD expertise in the UK, there were no significant UK supplier start-ups during the 1970s when US turnkey suppliers were establishing and dominating a CAD industry, and no significant diffusion of CAD into UK industry. CAD began to diffuse only from 1977 when US turnkey system suppliers began marketing seriously into the UK. This stimulated imitative entry by UK firms operating in specialised and low-specification market niches. Department of Industry (DoI) support for the turnkey installation at Baker Perkins in Peterborough had an important demonstration effect. In the absence of comparable UK products, it encouraged potential UK users to buy from US suppliers.

UK government policy has been unco-ordinated between supply side support and measures to promote adoption. Recent measures to subsidise CAD adoption (the DoI's CADCAM and CADMAT schemes) in the UK have tended to favour UK equipment, leading adopters to acquire rather specialised or low specification equipment. Simultaneously, policy has allowed small but technically strong UK CAD companies to be sold - in one case, by government itself - to US buyers. In UK policy, then, suppliers and users have not always been seen as related, while US policy linked and stimulated both.

CAD USE IN THE UK ENGINEERING INDUSTRY

The earlier development and diffusion of CAD in the US places potential UK users at a competitive disadvantage. But technology adoption involves more than simply buying a product off the shelf. It involves learning to exploit it - a learning fostered in the US aerospace industry by USAF activity. Equally, the creation of a pool of skilled manpower aids diffusion of technology between firms and industries. This pool was again, because of the type of policy pursued by government, relatively absent from the UK.

Blumberg and Gerwin(19) have argued that recent manufacturing technologies may be so far beyond the organisational capabilities of the US, UK and West German firms they studied (as distinct from Japanese firms) that they should settle for simpler methods. A study(20) at the Science Policy Research Unit (SPRU) of CAD adoption by 34 UK engineering firms suggests that barriers to CAD aquisition are deeply rooted in the skill structure of the firm at all levels, not least senior management.

Table 1 shows the motivations of firms sampled for adopting CAD. Many competed in world markets, and early adoption of CAD by US competitors forced adoption in the UK. Other factors were: the shortage of draughtsmen, which ended in about 1980 as recession cut engineering employment; and design complexity - notably in LSI microcircuits and complex printed circuit boards (PCBs).

CAD use can bring two major benefits:

-increased draughtsman productivity; and
-greater control over the design process.

Table 1: Motivations for involvement in CAD, by industrial sector

Motivation	Vehicles and Aerospace	Mechanical Engineering	Electrical and Electronic	Total
Design flexibility, complexity (includes design imperative)	-	5	4	9
Lead time, viability threat	4	4	1	9
Skill shortage, reduce dependence on contract draughtsmen	2	2	2	6
Experimental, unclear, other	4	5	1	10
TOTAL	10	16	8	34

Source: E. Arnold and P. Senker. Designing the Future: The Implications for CAD Interactive Graphics for Employment and Skills in the British Engineering Industry. Occasional Paper No. 9. Watford: Engineering Industry Training Board, 1982, p. 5.

Increased productivity reduces design costs, speeds documentation and reduces lead times. High productivity allows a firm to tender for more jobs, increasing its chances of getting work. The accuracy and presentation of tenders is improved, and more design work becomes possible at the tender stage, impressing potential customers. (Being able to show potential customers the CAD installation as proof that they are dealing with a 'high technology' supplier is also valuable. In at least two firms, this latter was probably the most important advantage gained in the early days of the installation.)

Productive use of CAD involved exploiting its line-drawing potential. (Indeed, obsession with payback sometimes led to a desire to raise drawing productivity in the short term, at the expense of reaping other benefits.) This implied setting up, and using, databases of standard component drawings. With this type of standards information on tap (rather than buried in a dusty standards manual), many users achieved greater volumes on bought-in components by maximising parts commonality.

Increased control over the design process was achieved in this way, through raising the level of accuracy of design, and by using one single database for the design as the 'master' version, ensuring that amendments and documentation were consistent. In the relatively small number of cases where CAD was electronically linked to NC machining, and to other types of computer-aided manufacturing (CAM), this tighter control of design extended through to manufacture. For example, the errors introduced by patternmakers 'blending by eye' were eliminated by NC of pattern geometry.

Most users encountered what we have dubbed the 'computerisation effect'; they found they had to reorganise themselves to cope with a computer, to make their procedures more explicit, and often to think for the first time in years about their procedures and standards. This brought improvements in the management and efficiency of the design process which could, in principle, have been reaped even without the CAD system. Some productivity gains normally attributed to CAD actually come via this computerisation effect.

Several users said that they only fully understood the benefits of CAD after they had bought a system. They had thought mostly about productivity before buying, and discovered the other aspects later. It normally took about a year from the time when CAD was discussed and investigated to the time when an order was placed. It then took on average about two years from delivery to bring the system up to 'best practice' productivity levels on 'live' production work. While learning was going on, it made no sense to buy more equipment, as this would only have increased the learning costs. Instead, firms waited until they achieved good performance levels before considering the purchase of further equipment.

The learning process in CAD adoption has two elements: operator learning; and management learning. Operator learning time can be drastically reduced by good training, plenty of access to the machine, and the development of appropriate component databases and instruction menus. Management learning is more complex and crucially affects the rate of operator learning.

The first management learning problem is system selection. Because there is no large body of user experience, managers generally do not know which are the relevant criteria for use in selection. By the time they find out they have been using a system for two years and are 'locked into' their supplier's goods by the effort they have invested in specialising instruction menus, archiving data in a particular format, and operator learning about how to operate the particular CAD system.

Industrial relations problems arise, if at all, normally when CAD is initially considered, so managers have to negotiate in ignorance of the most productive ways of running a system. Many, therefore, conceded single-shift working and 'non-dedicated' operation, and lived to rue the day. 'Non-dedicated' operation involves using CAD on demand (subject to a booking system), rather than full-time. It fits in well with the usual demand of the Technical and Supervisory Section of the Amalgamated Union of Engineering Workers (TASS), which organises draughtsmen, that _all_ drawing office staff should be trained in CAD, no matter how large the office or small the system. Most CAD systems were sufficiently 'user-unfriendly' that constant practice was needed to gain and maintain a high rate of output. As a result, 'dedicated' operators produced more than 'non-dedicated' ones, but they also tended to lose their engineering skills. One manager said that the CAD 'promotion' was the last his draughtsmen would ever get.

Some firms conceded small pay rises to buy themselves out of bad feeling over CAD. Others fought this tooth and nail. The sums involved were usually of the order of a couple of hundred pounds per person per year so these arguments were more about principle than about money.

Managers had little experience upon which to base cost-benefit analysis of CAD. Forecasts of benefits in potential users' investment appraisals were generally provided by salesmen and tempered with a little cynicism. Many managers felt their cost-justification was 'a bit of a fiddle', and were annoyed at having to present the benefits of the technology as if they were solely about increased drawing productivity. Unfortunately, this seemed to be the only language which many top managers could understand.

As a result, drawing office managers often played an ambiguous role. On the one hand, they had to justify CAD to top management in terms of savings in draughtsmen's wages. On the other hand, they had also to make a 'no redundancies' promise to the drawing office staff in the course of discussion and negotiation about CAD. Sometimes this was resolved by saving contract labour. Table 2 shows how firms in the sample justified acquiring CAD.

Table 2: Basis of cost-justification of CAD

Cost-justification basis	Number of establishments
Savings in drawing labour	21
Savings in drawing and design labour	3
Savings in estimators' labour	1
No cost-justification	5
No reason given	4
TOTAL	34

Source: E. Arnold and P. Senker. Designing the Future: The Implications for CAD Interactive Graphics for Employment and Skills in the British Engineering Industry. Occasional Paper No. 9. Watford: Engineering Industry Training Board, 1982, p.6.

Senker has argued that the prevailing US and UK management practice of treating individual departments of firms as profit centres and justifying capital investments on the basis of payback or, especially, discounted cash flow (DCF) criteria is inadequate to coping with new technologies such as CAD, CAD/CAM and flexible manufacturing systems(21). These offer systems gains, potentially improving the performance of the firm as a whole, not simply the one department which has to attempt a cost-justification. DCF and similar methods cannot take account of investment in the long-tern technological future. Hayes and Abernathy, similarly, argue that DCF and related approaches to 'management by numbers' have displaced engineering competence from corporate strategy formation, caused concentration on short-term investments, and promoted a downward trend in the productivity of the US economy, with managers making financially 'safe' investment decisions at the expense of more risky projects with larger but longer term payoffs(22).

POLICY IMPLICATIONS

UK policy has assumed that firms are perfectly rational entities capable of adjusting smoothly to technical change - that they conform to the neoclassical economic ideal of the firm - and that government need therefore only iron out imperfections in the availability of new technology and in information flows and help overcome capital barriers to technology acquisition.

In setting up the CADC as a government laboratory, the Ministry of Technology did not establish linkages between this source of new technology and potential users. As a result, CAD skills remained largely bottled up in the UK. In contrast, the USAF (wittingly or otherwise) forged such links.

Virtually all new technologies demand new skills of their users. The SPRU study of CAD suggested that the best practitioners of the old manual design and drawing skills are the most able to become good practitioners of the new computer-based ones. While engineers and managers cannot be trained for all possible future technologies, they can be trained to be better engineers and managers; to have an appropriate mix of engineering and management skills and an understanding of the interrelationship between these skills. The more skilled they are, the smaller the problems of new technology adoption.

Skill factors are crucial in the adoption of new technology. Management skills are needed in combination with engineering skills to assess not merely returns on investment at profit centres, but systems costs and benefits. Users must understand and analyse both the potential of computers and the complexities of their own work and practices in order to marry the two together. Presently, career managers rarely have engineering skills - indeed their training usually has more to do with accountancy and law (not that these are in themselves umimportant). Equally, engineers promoted into management are rarely trained in managerial and industrial relations skills.

In comparison with the managerial skills problem, the question of CAD operator skill is relatively trivial. CAD manufacturers offer draughtsman conversion courses which are at least partly adequate. It would help if draughtsmen gained some computer understanding - especially of the systems potential of computing - during their apprenticeship but they are already far better trained for the jobs that they do than most of their superiors. Thus, individual draughtsmen can learn CAD skills and attain best practice productivity levels in less than a year. Management, on the other hand, takes a total of about three years to do its learning: a year to make the investment decision and select a system; then two years or so to create an environment where operators can produce at best practice levels.

Managing a computer was often a new experience for drawing office managers. Some failed to put someone in charge of 'housekeeping' or to develop specialised menus. Some did not understand the importance of components databases for realising productivity gains. Almost all failed to put training or learning into their initial cost and time estimates. It took time to realise the benefits of the 'computerisation effect' and devise new working methods and standards. All these factors contributed to the learning period often being protracted.

While some users reached 'best practice' productivity levels in only 6 - 8 months, these were exceptional. At the other extreme, drawn-out industrial relations arguments continued for years. These latter cases were mostly in large but declining firms with poor industrial relations histories operating in mature industries. In other words, most of the industrial relations problems were not so much about CAD as about the more general state of things in the firm.

The SPRU study shows that technology adoption involves complex and lengthy learning processes. The benefits of even 'off the shelf' technologies cannot be gained without further ado. Technical change permits change in user firms, rather than enforcing such change(23).

An important strand in recent UK government policy has been the idea of 'awareness'. The take-up of microelectronics-based technologies in the UK has been seen as poor, and the policy response has essentially been an advertising campaign. This can, however, only be helpful to the extent that firms are already able to exploit the new technology. The same applies to subsidy schemes.

If engineers and managers in the UK were better trained, it is questionable whether there would be as much need for government policy to foster CAD adoption. Current UK policy is designed to foster adoption because the rate of CAD take-up us seen as slow and this tends to reduce UK competitiveness. CAD adoption is slow because of the skill and learning problems described here. Yet government policy is designed on the assumption that these problems do not exist.

Important policy conclusions follow. Current government policy is a laudable beginning, but in regarding CAD as a technological fix for an inherently perfect industrial system it fails to diagnose or tackle more serious problems. Clearly, it is important for government to promote the raising of management skill standards, both in the short run and through the longer term means of altering the way in which managers are educated and trained. Also, by German standards the UK workforce is severely under-trained(24).

In order to foster a successful CAD supply industry, government must relate technological development closely to adoption. Recent UK policy has seemed contradictory, with government simultaneously promoting mechanical and electronics CAD adoption (via separate and largely unco-ordinated schemes) and permitting the sale of the more promising small UK CAD companies (Shape Data Ltd, Cambridge Interactive Systems, and Compeda) to US interests. This makes it rather difficult to couple supplier and user development after the manner of the USAF.

Finally, although the importance of USAF activity has been stressed here, military activities are not the only ones capable of promoting the desired coupling. Indeed, since this beneficial commercial coupling occurred in the US largely as a by-product of the military's activities, it could be argued that the military route to advancing technology is prone to peculiar wastefulness.

The coupling between Intel as user and Calma as supplier of microelectronics CAD illustrates that other inter-firm routes are possible. A range of Japanese examples including machine tools and electronics CAD illustrates the potential of in-house coupling, where the large internal market of a diversified firm is exploited both as an inducement to,and a test-bed for, innovation. This route is presently precluded in UK companies which operate on a profit centre basis. In these companies, departments and subsidaries deal with each other at arms' length, so that managers have no means of sacrificing profits in one department to reap greater overall profit through the systems benefits of technologies such as CAD. Equally, government procurement (of all sorts) can provide important initial test-beds and markets if government is prepared to countenance the idea that its responsibilities go beyond obtaining a short run 'best buy' in procurement and match the role of many other governments in fostering new technological advances within the country. No doubt there is room for all these types of policy.

REFERENCES

1. The research upon which this paper is based was supported by the Engineering Industry Training Board. The opinions expressed here are, however, those of the author, and do not necessarily reflect the position of the Board.

2. PAVITT, K. (ed). Technical Innovation and British Economic Performance. London: MacMillan, 1980.

3. WIENER, J. English Culture and the Decline of the Industrial Spirit 1850-1980. Cambridge: Cambridge University Press, 1981.

4.GALBRAITH, J.K. The New Industrial State. Harmondsworth: Pelican, 1974, p.217.

5. SCHUMPETER, J.A. The Theory of Economic Development: An Inquiry into Profits, Capital, Credit, Interest and the Business Cycle. London: Oxford University Press, 1961.

6. SCHMOOKLER, J. Technical Change and the Law of Industrial Growth. in: GRILICHES, Z. and HURWICZ, L. (eds.) Jacob Schmookler, Patents, Invention, and Economic Change. Cambridge: Harvard University Press, 1972.

7. SCHMOOKLER, J. Technological Change and Economic Theory. in: ibid.

8. MYERS, S. and MARQUIS, D.G. Successful Industrial Innovation. Washington D.C.: National Science Foundation, 1969.

9. MOWERY, D. and ROSENBERG, N. The Influence of Market Demand upon Innovation: A Critical Review of Some Recent Empirical Studies. Research Policy, 8, 1979.

10. ROTHWELL, R., FREEMAN, C., HORSELEY, A., JERVIS,V.T.P. and TOWNSEND, J. SAPPHO Updated - Project SAPPHO Phase II. Research Policy, 3, 1974.

11. LANGRISH, J., GIBBONS, M.,EVANS, W.G., and JEVONS, F.R. Wealth from Knowledge: Studies of Innovation in Industry. London: MacMillan, 1972.

12. FREEMAN, C. The Economics of Industrial Innovation. Harmondsworth: Penguin, 1974, p.166.

13. WASSON, L. How Predictable are Fashion and Other Product Life Cycles? Journal of Marketing, 32, 1968.

14. ARNOLD, E. Information Technology in the Home: The Failure of Prestel. in: BJORN-ANDERSEN, N., EARL, M., HOLST, O. and MUMFORD, E. (eds.) Information Society: For Richer, For Poorer. Amsterdam: North-Holland, 1982.

15. COONS,S.A. An Outline of the Requirements of a Computer-Aided Design System. in: AFIPS Conference Proceedings Vol 33, 1963 Spring Joint Computer Conference. Santa Monica, California: American Federation of Information Processing Societies, 1963.

16. SUTHERLAND, I.E. Sketchpad: a Man-Machine Graphical Communication System. in: ibid.

17. SHIMSHONI, D. The Mobile Scientist in the American Instruments Industry. Minerva, 8, 1970.

18. RADLEY, D. Computer Aided Design in the USA. Computer Aided Design, 1, 1969.

19. BLUMBERG, M. and GERWIN, D. Coping with Advanced Manufactruring Technology. Berlin: International Institute of Management, 1981.

20. ARNOLD, E. and SENKER, P. Designing the Future: The Implications of CAD Interactive Graphics for Employment and Skills in the British Engineering Industry. Occasional Paper No. 9. Watford: Engineering Industry Training Board, 1982.

21. SENKER, P. Some Problems in Justifying CAD/CAM. Paper presented at AUTOMAN '83 Conference, Birmingham, May 1983.

22. HAYES, R.H. and ABERNATHY,W.J. Managing our Way to Economic Decline. Harvard Business Review, July-August 1980.

23. BELL, R.M. Changing Technology and Manpower Requirements in the Engineering Industry. Engineering Industry Training Board Research Report No.3. Brighton: Sussex University Press in Association with EITB, 1972.

24. PRAIS, S.J. Vocational Qualifications of the Labour Force in Britain and Germany. National Institute Economic Review, October 1981.

Robotics in manufacturing organisations

James Fleck

INTRODUCTION

In this chapter, I will discuss some of the issues facing management in the introduction of industrial robots. The points raised have been drawn from a two-year project on the diffusion of robots into British manufacturing industry, carried out in the Technology Policy Unit, at the University of Aston in Birmingham(1).

This project took place against previous work started in 1976 on the development and diffusion of industrial robots(2). Current work is going on to look at the influence of national strategies and policies on the diffusion of robots, and at ways of effectively modelling the robot adoption process(3). In the background were other research projects on: the impact of microelectronics in general; management attitudes towards innovation; and union-management agreements over the introduction of new technologies. In addition, doctoral studies of the introduction of machine tools; national policies towards computers and machine tools; and automation in the car industry also provided relevant input(4).

We investigated 32 cases of robot introduction, involving some 137 robots and more than 100 of the simpler pick-and-place devices, as well as several cases of potential users, and non-users. But, before discussing the details of the case studies, a brief examination of what robots are is in order.

WHAT ARE INDUSTRIAL ROBOTS?

By now there are few people who have not seen a robot in action on television, or even in the flesh (or rather metal!). Quite clearly industrial robots are a far cry from R2-D2 of 'Star Wars'. The standard configuration is that of a single arm reaching out into unconstrained three-dimensional space. It is the unconstrained aspect that gives robots their lifelike character when in motion (quite unlike most machines) and, more importantly, that gives them their special flexibility and relative freedom to be interfaced with any other sort of machine tool or task. Thus, typically, one finds industrial robots in a variety of manufacturing processes, from paint spraying and spot welding, diecasting and other machine unloading, to packing, palletising, and other more intricate tasks such as simple assembly(5). The applications covered by our case study examples are indicated in Table 1.

Table 1: Breakdown of sample and other surveys by broad category of application

Application / Survey	Loading/ Unloading (%)	Manipulation and Process (%)	Joining (%)	Surface Treatment (%)	Assembly (%)	Other/ Unknown (%)	Total
TPU, 1979	49 (30)	7 (4)	58 (35)	32 (20)	4 (2)	14 (9)	164
BRA, 1980	147 (40)	17 (5)	107 (29)	69 (18)	5 (1)	26 (7)	371
SAMPLE (1978-82)	54 (39)	7 (5)	32 (23)	26 (19)	2 (1.5)	16 (12)	137
BRA, 1981	202 (28)	23 (3)	254 (36)	88 (12)	15 (2)	131 (18)	713

The point I should like to emphasise here, however, and one that is often overlooked, is that there is a wide range of robotic devices, with a variety of operating characteristics - different power systems, different structural configurations, different control systems, different load capacities, and so on. There is a continuum from the very simple pick-and-place devices, costing only a few thousand pounds and not regarded as true robots in many definitions(6), to extremely sophisticated computer-controlled devices costing more than fifty thousand pounds. The simplest pick and place devices have a restricted range of movements, with severe limitations on the extent to which they can be programmed. Programming is generally achieved by means of screw-adjusted mechanical stops for setting the range of movements, and electromechanical or pneumatic switches for selecting the desired sequence which may include only a few steps. With the more sophisticated devices recognised as 'true' industrial robots, the range of movements is far greater, while electronic programming enables extremely complex and variable sequences of movements to be attained.

This variety and range, then, offers the first challenge to managements intending to use robots. Robots may have an element of flexibility, some general purpose characteristics, but this does not mean any old robot can do any task. The most successful robot installations are those in which the robot is well adapted to the task. And with the development of new applications, more specialised robots are coming onto the market all the time, along with different approaches to robotisation such as modular robots and mobile robots(7). So the first step in the successful introduction of robots must be the securing of a reliable, competent and knowledgeable guide for selecting the most appropriate system.

WHAT HAVE ROBOTS GOT TO DO WITH IT?

There are several answers to this question. Firstly, the most pragmatic answer is that promotional programmes for robots are part and parcel of the IT initiatives, and free or subsidised advice and consultancy is available from bodies such as PERA (Production Engineering Research Association)(8).

Secondly, and more fundamentally, industrial robots are not stand-alone devices which can be bought off-the-shelf and plugged in immediately to a chosen application. The robot forms only part of an overall system, even in the simplest applications such as die-casting, and the interface between the robot and the other machines is crucial. This interface consists primarily of interlocks that ensure the different devices mesh together correctly and the whole system operates safely - in short the patterns of information flow that direct how the technology operates. There are emerging, albeit some way off at present,integrated manufacturing systems comprising more or less conventional automation equipment along with modern, numerically controlled machines, serviced and physically integrated by industrial robots, under the overall control of the computer-based technologies more readily recognisable as information technology.

One particularly impressive example of this line of development is the ZF (Zahnradfabrik Friedrichshafen AG) flexible manufacturing system for the machining of metal parts, installed in their factory in Southern Germany. This consists of 14 machine tools, 15 robots, and a loading gantry all controlled by a mainframe computer. The machines are laid out in two lines, with the gantry robot running down the middle storage space. Each machine tool works independently, served by a robot from buffer stores held in the central area, and the system will produce a wide variety of spur gears. But it has to be stressed that all the components are necessary, and deserve consideration under the information technology heading, including of course robotics.

Finally, as I hope will become clear, a considerable amount of effort, and therefore commitment, is required to achieve a successful robotic installation. The key here is appropriate advice and skills, the knowhow of competent people - in short information, albeit information embodied in people.

A LEADING QUESTION

I want to be provocative here, by asking whether the great concern generated over robots during the last few years is really justified(9). Are robots really something special, totally distinct and unique in themselves? Or are they, as I have already suggested, just another part of automation, one slice out of a continuum of automatic devices, and only one sub-component of modern manufacturing systems? This question is worth bearing in mind, I think, to enable one to more readily and critically assess the promotional literature, of which there is a huge amount. A healthy dose of scepticism in reviewing this literature is well-advised and will aid in the robot selection process.

Results from an analysis of case study material are presented to illustrate the following discussion. The work was essentially explorative of the structure of the robot adoption process, rather than being directed towards the testing of hypotheses deriving from well-formed theories of innovation, which in any case do not yet exist. Consequently, it was not considered appropriate to apply more sophisticated procedures in the analysis of the case study material than simple percentage breakdowns to provide comparisons between the successful and unsuccessful cases. In the original analyses a trichotomous breakdown was used: factor certainly present; certainly not present; and the situation uncertain or unknown. But for presentation purposes in this chapter we have only given the frequencies in which the factor under consideration certainly was present, for both the successful and unsuccessful sub-samples(10).

MOTIVES FOR INTRODUCING ROBOTS

In a third (32%) of the cases examined, robots were seen as a solution to specific, already identified, problems in the production process. In 19% of cases, robots were used in emulation of similar applications seen elsewhere. In a further 19% of cases, however, an application was found only after management had already decided to invest in robots, often for prestige reasons. The situation was not clear in the remaining cases.

PATTERNS OF SUCCESS AND FAILURE

The introduction of robots, however, did not always result in simple success. (Our criterion here for success was the use of robots in actual production). In well over a third (44%) of our cases, initial failure was experienced, and half (22% of the total) of those abandoned robots altogether. In some cases (6%), failure was followed by continuing further development, while in another 16% of cases initial failure was followed by eventual successful installations - three-quarters of which used far simpler robotic systems (in some cases little more than pick-and place devices), better suited to the particular application, or cheaper. Just over half of the cases (56%) experienced straight-forward success to varying degrees.

Moreover, subsequent monitoring of the firms covered by the study has revealed that several making successful use of robots have gone into receivership, or otherwise met serious financial difficulties on account of the current recession. This very clearly illustrates that robots are not a panacea. The survival of a firm requires good management in all respects as well as the efficent implementation of new industrial processes. The use of robots requires good management and will not excuse bad management; automation and liquidation are not necessarily mutually exclusive possibilities(11).

THE ADVANTAGES OF USING ROBOTS

The outstanding advantage of robots is that managerial and technical control is enhanced. An improvement in process control was found in more than half of the successful cases of robot adoption, and was less in evidence among unsuccessful users. Improvements in quality and consistency were also found more often in successful than unsuccessful cases, but were not always reported, however.

Table 2: The advantages of using robots

Factor	Successful cases (%)	Unsuccessful cases (%)
better process control	52	25
improvements in quality	35	12
improvements in consistency	43	12
elimination of human variability	61	12
labour displacement (for robots, not pick and place devices)	61	-

The better process control given by robots is largely tied up with the elimination of human variability in the production process. As well as yielding better consistency and quality, this reduced the dependence upon human labour (by displacing labour) and the vulnerability of the processes to possible strike action. The elimination of labour, or labour saving, provides the prime economic motivation for introducing robots. The actual figures for our case studies proved difficult to ascertain and depended upon the particular application, involving such factors as: the technology itself; the level of automation before and after robotisation; the extent of shift working (which in turn depends on market conditions, of course); and above all the degree of integration with other technologies into a complete system. For instance, one system with an estimated labour displacement of 23 people per shift comprised one robot, a minicomputer, and supporting special purpose automation - how therefore should the labour displacing effects be allocated between the components?

However, we found that the average net effects were in the region of 2.5 people per robot, taking into account patterns of shift working and increases in indirect labour requirements due to the management, engineering, and maintenance effort demanded. These indirect requirements were largely concentrated into the early installation phase before normal productive use was achieved.

CONDITIONS FAVOURING SUCCESS IN THE USE OF THE ROBOTS

Table 3: Favourable conditions for successful robot adoption

Factor	Successful cases (%)	Unsuccessful cases (%)
presence of existing automation	83	25
previous experience of automation	87	38
bad working conditions (dirty, hazardous, or tedious)	74	50
labour problems (absenteeism, turnover, shortage)	57	38
complex approach to feasibility employed	17	0
other systems considered	61	25
complex approach to *post hoc* assessment employed	13	0
problems over the availability of capital experienced	31	13

Given that not all instances of robot adoption were equally successful it is worth looking at those conditions that appeared to be favourable for robotisation. The clearest such factor was the presence of existing automation in the adopting firm, found in 83% of the successful cases and only a quarter of the unsuccessful cases. As well as providing an opportunity for robotisation on the technical level, in the form of robots servicing existing automatic machines, this was important in organisational terms. Previous experience of automation provided management and engineering personnel with some basis for knowing what to expect from further automation in the form of robots, and also implied the actual existence in the first place of appropriately competent people.

Another condition clearly favourable for successful robotisation was the presence of bad working conditions in the workplace to be robotised. Improvements were made in terms of dirty or hazardous conditions rather than reduced monotony. In some cases, on the contrary, monotony was worsened due to increased machine pacing, and more intense levels of working. For example, in one case of injection moulding machine unloading, a single robot serviced two machines, with the output channelled to one human worker to desprue and pack the mouldings. Output rates were increased and cycle times more constant. Previously, one worker had been responsible for the entire operation of one machine: unloading, inspecting, resetting, despruing and packing, with, of course, full control over the cycle. The job then had greater variety and also a lower intensity due to waiting time during the injection phase.

Bad working conditions were often associated with the existence of labour problems - absenteeism, turnover and even shortages - and this proved to be another facilitating condition. But it is clear that robots will not always be restricted to such dirty or problematic areas. Assembly tasks in particular are increasingly seen as an area for robotisation, and these are often clean and desirable jobs. The social milieu of work can often overcome tedium; indeed, in some cases a regularity in the work task can facilitate social interaction as the workers can carry out the task without much thought. One large manufacturer we talked to had reverted from 'enriched' workstations, in which each worker built a complete sub-assembly, to the old style, one component added per worker assembly lines for this reason (this incidentally was not one of our robotic cases).

Careful attention paid to the economics of the robotic installation, with explicit consideration given to alternative systems - either using simple devices or non-robotic solutions - was associated with the successful installations more often than the unsuccessful ones. Such attention ideally should extend to the analysis of the running of the robotic system after its installation as well, as otherwise it is often not clear whether the system performance is in fact all that it was expected to be. It was surprising how rarely such *post hoc* assessment was carried out, and in general we found a distinct lack of systematic and careful consideration using techniques more sophisticated than pay back on labour savings - despite the wide discussion of such sophisticated methods in the literature(12). But in the relatively few cases where such techniques were employed, the installations were successful. This is perhaps worth emphasising as there is wide support for the belief that use of such techniques can hinder the uptake of sophisticated technology and is therefore not desirable. The 'ignorance is bliss' approach will undoubtedly speed up the diffusion of robots, but there is a danger that it will also encourage less than the best use of the new technology, and therefore tie up capital in fashionable but not cost effective projects. Ironically, though our information on the availability of capital for robotic projects was limited, from the information we did get it appeared that the unsuccessful cases had fewer problems in getting finance. This suggests that restriction on capital may force a closer scrutiny of investment, leading to better projects eventually getting funded.

PROBLEMS OVER THE USE OF ROBOTS

As already mentioned, the installation of robots is not merely a matter of just plugging in an off-the-shelf model. As with any sophisticated technology, problems can and do occur. Technical inadequacies did appear to be a source of problems, especially in the less successful cases and especially with the earlier models. On the

Table 4: Problems over the introduction of robots

Factor	Successful cases (%)	Unsuccessful cases (%)
robot inadequacy or unreliability	61	87
system problems	61	50
long development periods (>1 year)	52	75
lack of experience of automation	0	25
considerable managerial effort involved	52	75
organisational resistance (all sources except labour)	30	63
provision of maintenance labour was a problem	60	63
provision for training was required	61	63

other hand, the reliability of later models was specifically commented on in several cases as well. Technical difficulties were also experienced at the systems level, in getting the overall robot plus peripherals plus control plus other machines to work, but these appeared to be more in the nature of challenges to be overcome, rather than insuperable problems leading to failure.

The installations, even relatively simple ones such as plastic injection moulding machine unloading, typically took a long time to work out and debug - in surprisingly many cases (19%) more than two years. This is perhaps only to be expected at this relatively early stage of diffusion but it has to be borne in mind and allowance made. In some of the cases of failure it was possible that the adopting firm had simply not put enough effort into the installation. Seduced by the idea of easily implemented general purpose robots, they had given up too early. These were often companies without any extensive experience of automation.

Associated with these long development periods, it was evident that considerable organisational effort was required to get installations working - in other words, an organisational learning period was required. It was not possible to identify any single common problem that caused long times to be taken, rather a whole host of things had to be sorted out - interdepartmental co-operation; technical debugging; safety checking; and organisational resistance of one form or another.

In some cases, for instance, a shift was required from an organisational structure appropriate for the management of people, to one appropriate for the management of machines, requiring a corresponding shift from inter-personal skills to engineering skills. Robots do not respond to the kind of persuasion that people do - 'I'll see you alright if you get this lot out by 4pm, Fred'. This shift can take time to happen, and one can find day-to-day production pressures leading to a robotic system being switched off in the event of any problems and the manual system being re-instated, instead of the fault being diagnosed and put right. Ironically, the good reliability of the robot can exacerbate this tendency due to the resulting lack of practice in correcting faults. Some programme of preventive maintenance is probably the best way of avoiding these difficulties.

In other cases, a lack of technical expertise on the part of management was pointed to - a problem often identified as a key British failing by some analysts(13). This can lead to unrealistic demands being made for the robot performance and the time taken to install the new systems. On the other hand, engineering personnel, who do have the technical expertise, may have strong personal commitments to the existing systems which they themselves perhaps had installed. Consequently, they may resist robotisation, and see potential problems as looming very large indeed.

Such resistance can be further exacerbated by departmental boundaries and disciplinary divisions. The general lack of a sophisticated approach to the economic assessment of robot systems (mentioned above), for instance, appears to be associated with the oft noted division of expertise between the separate professional groups of engineers and accountants. Robotisation involves a shift to more capital intensive production, with a resulting increased need for the improved utilisation of such expensive machinery. Accurate economic assessment in these cases clearly requires a combination of technical and financial expertise, and this does not appear to be widely available at present. But these and other forms of organisational resistance, though marked in a few instances and more prevalent in the less successful cases of robot adoption, were not common.

In no case did we find evidence of resistance on the part of labour, despite a general lack of consultation at the very early stages of an installation. In general the pattern appeared to be one in which the particular installation was developed (often under a shroud of secrecy), until ready for productive use, and only then was the workforce informed. Under such conditions it was perhaps surprising that more trouble was not experienced. Indeed it appears that the fear of labour resistance in the UK has been responsible for slowing down the diffusion of robots to a greater extent than any such resistance itself. This is undoubtedly one of the factors underlying the peculiarly British pattern of robot installations in ones or twos across many companies, with relatively few large users (14).

The greatest single problem commented on, however, was the difficulty of retaining suitably trained maintenance men, especially in firms without technical personnel due to a lack of any other automation. Some form of training for programming and maintenance for robots was required in nearly all our cases of robot introduction. On a broader view, the diffusion of robots in UK manufacturing industry is certainly hindered by a shortage of competent and skilled men, not only at the technician level, but also at the general engineering and consultancy levels. This is partly a consequence of the current stage of diffusion: there simply have not been enough robots around for sufficient time to permit many people to gain the requisite experience to be deemed truly competent. But it is undoubtedly also the result of endemic weaknesses in the British system for technical education and training.

This returns us to my starting point where I advocated that management should seek competent advice to help them in selecting a robot: clearly great care must be taken also over the source of advice, given the shortage of experienced robotics personnel. If possible, a suitable person within the company should undertake to learn all one can about robots and robotisation, and the sooner the better. If someone is being brought in from outside to take care of the introduction of robots, great care must be taken in selection. Credentials should be viewed sceptically - ask how many robots the candidate has installed, in what applications, and ask to see them. The great publicity surrounding robots and the public money available has brought out all manner of 'cowboy consultants' and instant experts.

CONCLUSION

To conclude and provide an answer to my leading question: I do not think robots are so unique or so different that they warrant all the fashionable attention they are receiving. Less talk and more solidly based action is required to improve Britain's industrial performance: there is a danger that robots are becoming an international virility symbol.

Robots are not a panacea, but are merely one form of automation among others. They require great care in their implementation as do all technologies, in order to get the best out of them. The sorts of advantages and problems thrown up by robots are not so different from those found, for example, in the adoption of numerically controlled machine-tools; or in conventional special purpose automation itself. Indeed many of the issues raised by robots can be found in writings from the great automation debate in the 1950s and 1960s. In particular, the shortage of appropriately trained manpower was pointed out in the 1956 Department of Scientific and Industrial Research report on automation, as was the need for special training and education to eliminate that shortage:

> 'The shortage of trained managers, engineers and technicians may be the most decisive factor (in the adoption of automation). It arises because the technical complexity of industrial production is increasing, partly through automation and there are not enough scientists and engineers to go round. This problem is already so important that a major re-appraisal of university and vocational training in this country may be necessary'(15).

Yet nearly thirty years later the same problems are still with us. Why?

Information Technology Year has certainly increased general awareness of the cluster of new technologies now available for use in the factory and the office. However, the emphasis on the technology part of IT has perhaps created, or at least reinforced, the attractive notion that problems over the flow and use of information can be solved by buying in more new gadgetry. With Information Technology Year behind us, we should now turn our attention to providing the appropriate knowledge and expertise for making effective use of the technologies becoming available. The question is not merely one of information <u>technology</u>, but of information embodied in <u>people</u>. All the information in the world is of little use unless we have competent and skilled people to use it and give it meaning.

NOTES AND REFERENCES

1. This project, entitled 'The Diffusion of Robots in British Manufacturing Industry', was funded by the Joint SERC/SSRC Committee. This paper is based on results given in the Final Report from that

project (TPU, June 1982). A review of the literature, informed by this project, is in preparation: FLECK, J. The Introduction of Industrial Robots. London: Frances Pinter, forthcoming.

2. See ZERMENO-GONZALEZ, R. The Development and Diffusion of Industrial Robots. Ph.D. 1980, The University of Aston in Birmingham.

3. This new project, 'The Adoption and Diffusion of Industrial Robots' started in Oct. 1982 and is funded by the Leverhulme Trust.

4. For further details of this work and resulting publications, see Current Research Activities, TPU, University of Aston in Birmingham.

5. An excellent review of the range of robot applications is given in ENGELBERGER, J.F. Robotics in Practice, London: Kogan Page, 1980.

6. For example the British Robot Association (BRA) definition: 'An industrial robot is a reprogrammable device designed to both manipulate and transport parts, tools, or specialised manufacturing implements through variable programmed motions for the performance of specific manufacturing tasks.'

7. A review of the contents of any of the major conferences in the area e.g. the International Symposia on Industrial Robots (ISIR), will illustrate this point.

8. This service is sponsored by the Department of Industry; information on The Robot Advisory Service is available from PERA, Melton Mowbray, Leicestershire, LE13 0PB.

9. The growth of this can be dated to 1977 when the BBC Horizon Programme 'The Chips are Down' put robots (and microelectronics in general) firmly on the public agenda.

10. Some 45 cases in all were studied. Of these only 32 actually attempted to introduce robots. 24 were ultimately successful, while 8 were unsuccessful. One of the 24 (in fact the car industry case) was not covered in the percentage breakdown analysis, as it was very much a special case.

11. 'Automate or liquidate' is the widely quoted slogan attributed to Kenneth Baker, Minister for Information Technology.

12. See, for example BENEDETTI, M. The economics of robots in industrial applications. The Industrial Robot, 4 (3), 1977, pp.109-118; or 'Robot Economics', ch. 7 of Engelberger's book, op cit.

13. For example, JOLLY, B.S. New technology demands a new breed of engineer-manager. Electrical Review, 207 (4), July 1980, pp. 27-28.

14. This distribution is often commented upon and sometimes seen as a good thing in that it implies a distribution over many companies of experience in introducing and using robots. This is perhaps true, but

it also implies that existing expertise is being rather thinly spread, which is not such a good thing.

15. DEPARTMENT OF SCIENTIFIC AND INDUSTRIAL RESEARCH (DSIR). Automation: a report on the technical trends and their impact on management and labour. London: HMSO, 1956, p.54.

Organisation design for CAD/CAM

Graham Winch

INTRODUCTION

Although the use of the main-frame computers for sales ledger and inventory control functions was fairly well established, it is only during the mid to late seventies that the use of information technology for major design and production functions penetrated batch manufacture in the metalworking industries. The nomenclature of the new technology is not consistent, but for the purposes of this chapter, the following definitions will be used. Computer Aided Design is the use of computers for the generation and analysis of design concepts. Computer Aided Draughting (CAD) is the use of a computer for the translation of design concepts or sketches into drawings which inform manufacturing how the product is to be made. Computer Aided Manufacture (CAM) is the use of computers for the programming of numerically controlled (NC) machine tools and the planning of production processes. Computer numerical control (CNC) is a refinement of NC involving the computerisation of the machine tool controller(1).

The use of computers for design takes many forms, and need not have any immediate implications for the rest of the organisation. However, the implementation of CAD and CAM systems, and CNC machine tools, is having a major impact in the metalworking industries at the beginning of the eighties. In particular, the linking of CAD and CAM to form CAD/CAM systems, on the one hand, and the linking of CNC machine tools with automated handling devices, often robots, to form flexible manufacturing systems (FMS) on the other have major implications for the organisation structure of the firms concerned. The work reported here is based on in-depth case studies conducted in four batch engineering firms on five sites. All cases were part of large corporations, but acted as independent profit centres responsible for their own capital investment policy. They all manufactured capital goods to order with average batch size in single figures on sites employing over 1000 workers (see Table 1).

Table 1		
	Case A	press tools
	Case B	electrical machines
	Case C	food processing and other machinery
	Case D	large electrical machines (two sites)

THE PRODUCTION PROCESS

The task of any capitalist manufacturing unit is to sell products at a satisfactory return on capital employed. In order to accomplish successfully this task, such organisations must process both information and materials(2). The manufacturing process is initiated when sales receive a firm order from the customer. Information processing in small batch metalworking can then be divided into four distinct stages, in which the characteristics of the product are specified at increasingly greater levels of detail:

1. Design - the overall conception of the product has to be outlined and clarified, and then put into a communicable form.

2. Draughting - the detailed characteristics of the product must be specified in order to eliminate ambiguity once material processing starts.

3. Planning - the sequence of material processing must be specified, and costings calculated. The availability of raw materials and production resources such as jigs and fixtures, and NC programs must also be established.

4. Production control - once material processing is under way, its progress must be monitored against the criteria laid down in the previous three stages.

In conventional parlance, the 'engineering' side of a batch metalworking firm processes stages 1 and 2, while the 'manufacturing' side processes stages 3 and 4 as well as materials. However, it is not unusual for stage 3 to be located on the engineering side. In case B, production planning was moved from engineering to manufacturing during the course of the research.

After the planning stage, material processing starts and proceeds from the preparation of the raw materials by processes such as founding and fabrication, through the manufacture of components by techniques such as machining and pressing, and on to final assembly in the fitting shop. Throughout these processes, progress is monitored and reported back to the production control function, and production is re-planned if necessary. The levels of complexity and uncertainty involved in batch production create a distinctive industrial environment(3). Although it is clearly materials processing which is the key dominant task of a manufacturing organisation, most of the issues and problems of organisation structure derive from the information processing stages. It is these issues which this chapter will examine.

The body of literature that will be drawn upon for the analysis of organisation structure can be grouped together as 'contingency theory'. Its central tenet is that organisations design their structures to cope with the various contingencies of technology, market environment, size, and so on, that they face. Those organisations closest to the optimal structure for their particular set of contingencies will be the most successful economically. Thus the literature is based upon a premise of

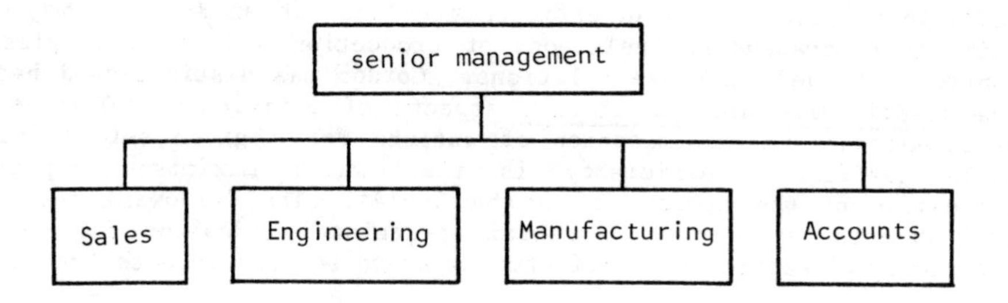

FUNCTIONAL ORGANISATION

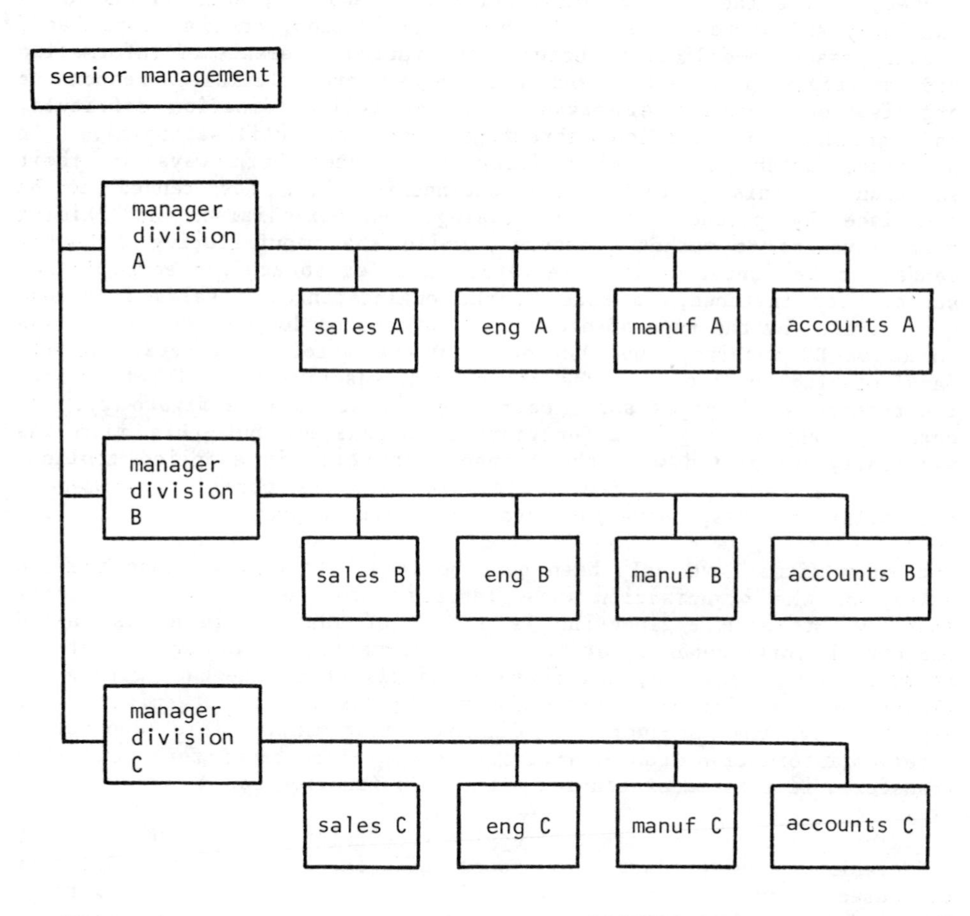

Figure 1 PRODUCT ORGANISATION

'efficiency', but it is an efficiency defined in an a-social way which fails to comprehend the relations of production and the struggles for control embedded in those relations. Gordon has distinguished between the quantitative and qualitative aspects of efficiency. Quantitative efficiency is the maximisation of outputs for a given set of inputs, while qualitative efficiency is that which maximises capitalist domination of the process of production(4). The following discussion will show the ways in which certain organisation designs increase the efficiency of capitalist manufacturing organisations in both senses.

ORGANISATION STRUCTURE IN BATCH ENGINEERING

A perennial question of organisation design is whether to structure by function or product (see figure 1). Functional organisation, it is argued, allows the maximum utilisation of resources, such as expensive machinery and scarce skills. Product organisation, on the other hand, greatly eases co-ordination between the various elements of information and materials processing, and encourages more flexible, responsive organisation. Functional organisation makes co-ordination difficult, and product organisation threatens resource utilisation(5). In practice, batch metalworking firms have faced both ways in their solution to this problem. The engineering side has tended to be organised by product, thereby easing customer liaison and aiding responsiveness to market pressures, while the manufacturing side has tended to be functionally orientated in order to achieve economies of scale. For instance, in case B, the engineering and sales functions were divided by the type of electrical machine produced - whether it was an AC or DC machine, and whether is was a motor or generator. The manufacturing function, on the other hand, was divided by function into the machine shop, press shop, coil shop and so on (see figure 2). In case A, engineering was functionally organised, but this firm was vertically integrated with the company for which it supplied tooling, and only a small proportion of its work was for outside customers - flexibility and responsiveness were not at such a premium.

Such firms have only been able to face both ways because the two halves of the organisation were largely uncoupled. The traditional workflow in batch engineering is a form of what Thompson has called sequential inter-dependence(6). The information flows and materials flows follow a sequence, and there is little or no feedback within the flows(7). The liaison between the two halves of the organisation is achieved by the production planning department. This department receives information from engineering in a product based form, and then transforms it into manufacturing plans in a function based form.

The practice is for the engineering side to 'issue' product information to the production planning function. The production plan is then used to initiate production activities, but not to control them. The production plan is continually amended by those on the shop floor, to meet the daily requirements of resource utilisation and delivery dates(8). The re-allocation of production priorities, so far as delivery is concerned, is the job of the progress chaser who expedites certain elements of materials processing in order to meet the

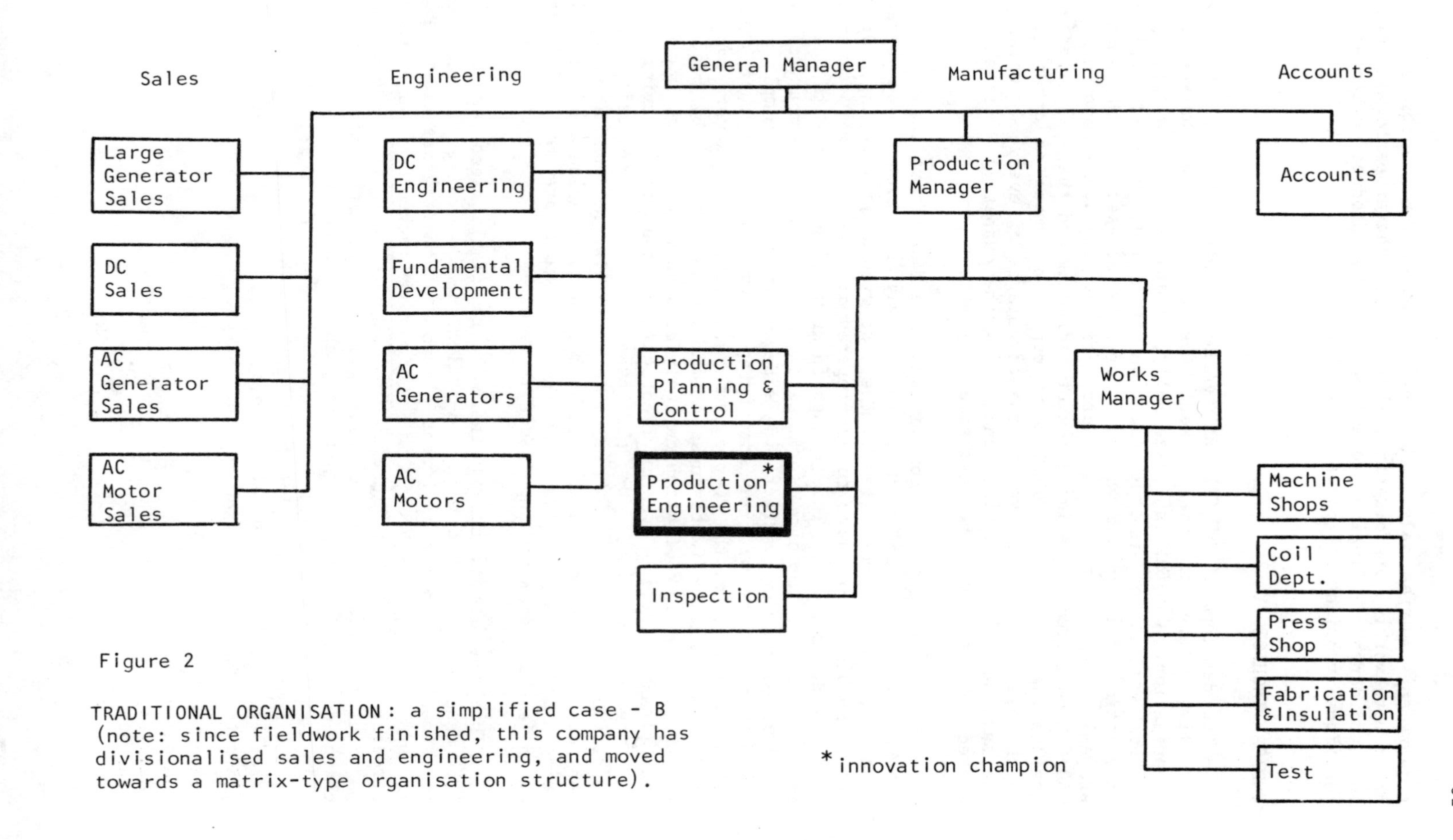

Figure 2

TRADITIONAL ORGANISATION: a simplified case - B (note: since fieldwork finished, this company has divisionalised sales and engineering, and moved towards a matrix-type organisation structure).

requirements of the sales function. Thus in the traditional form of batch metalworking organisation, the progress chaser performs a crucial, but rudimentary, linking role between the product orientated and functionally orientated parts of the organisation.

CAD/CAM TECHNOLOGY

The development of CAD/CAM technology has important implications for the nature of the linkages between the two halves of the organisation. CAD/CAM is, essentially, a data base technology. The key to its utilisation is the creation of a data base of engineering information at stage one which can then be accessed at the subsequent stages. Most of the direct productivity improvements reported for CAD/CAM come from the elimination of redrawing and the ease of access to engineering standards. ACARD also report a number of other benefits which essentially boil down to the advantages of everybody working off the same set of information, and having the output of their activities stored and ready for others to access(9).

The development of a common data base means that those who access it downstream have an interest in the way in which it is generated upstream. This is most clearly the case with NC programming. Drawings for NC machine tools have to be prepared in a certain way, dimensioned from one datum point. This change in draughting practice is due to the way in which NC machine tools are programmed. A number of those interviewed on the manufacturing side of organisations complained that the drawing office was not specifying datum points and that this was making more work for the NC programmers. There are, however, many other ways in which the effectiveness of CAM can be improved by planning at the CAD stage, many of which can only be discovered by trial and error. Such planning implies greater feedback from the manufacturing side to the engineering side. On the other hand, the necessity to draw specifically for NC, and often for a particular NC machine tool, means that manufacturing decisions such as machine allocation are increasingly being taken by engineering. If such decisions are to be effective, again feedback from the shop floor is required.

Thompson has described these sorts of linkages as reciprocal interdependence(10), in which co-ordination is achieved through mutual adjustment between the different elements of the organisation. To put it another way, CAD/CAM is an integrating technology which requires stronger organisational linkages for its effective use. These changes particularly affect stages two and three of information processing. The change described here is a shift of emphasis rather than an absolute one - in comparison with the mass production of metal goods, even the most traditional batch firm already displays a high degree of reciprocal interdependence(11). The shift is, however, of such a magnitude as to generate considerable organisational stress in many cases.

VALUE ENGINEERING

CAD/CAM is not, however, the only development that is encouraging integration in batch metalworking. A study of two comparable textile machinery factories showed that, despite greater investment and higher worker effort in the British plant, the American one was more efficient. This was put down mainly to the British engineering. 'The tendency in Britain is to design something and then hand the design to the production side and let them make it. In the US something is designed and then costed and then redesigned to lower the cost, and this might be done several times over before a design is finally adopted'(12). The result of this effort was that one type of machine in the US required 70% of the direct labour required in Britain, and for a second type, the American labour requirement was only 48% of that in Britain.

This technique of costing design options from a manufacturing point of view is known as value engineering(13). Effective design for manufacture requires feedback from manufacturing to engineering because it is against previous manufacturing experience that design options are evaluated. Again, there is a shift towards reciprocal interdependence in information processing and the necessity for stronger organisational linkages, in particular between stages one and three.

CAD/CAM and value engineering are not totally separate developments. Although value engineering does not necessarily imply capital investment, the effective utilisation of NC and FMS on the shop floor does necessitate a value engineering approach. These technologies provide new opportunities and pose new constraints for manufacturing techniques, and so feedback on manufacturing experience becomes more and more essential as investment proceeds on the shop floor.

ORGANISATION DESIGN FOR CAD AND CAM

The implications of NC for organisation structure were appreciated in the sixties, but the impact was largely restricted to the manufacturing side(14). These trends were readily apparent in the cases studied, and in general involved the removal of decisions about machine and labour utilisation from the shop floor to the planning office. The results of this for the skilled machinist have been examined in detail and characterised as _deskilling_(15), but the impact on the supervisor is equally severe, as key decisions about the allocation of resources are removed from the shop floor. Both these trends have been under way since at least the beginning of the century and NC and CAM are only extra elements in a changing scene(16). A similar process appears to be occuring on the engineering side, but it is much less well documented. Cooley has argued that CAD deskills drawing staff(17), and it is certainly true that CAD means that staff at stage two have less opportunity to stamp their own identity on drawings. From the case study evidence, it also seems to be the case that the designers are doing more of the detailed specification of the product at stage one when inputing to the data base. Again, as Cooley himself notes, the fragmentation of the draughtsman's job has been happening since the thirties, and is linked with the increasing recruitment of graduate engineers into design stifling promotion from amongst the ranks of draughtsmen(18). While engineering decisions are being increasingly taken at stage one rather than stage two, this does not appear to have created significant organisational tensions. In none of the cases had any organisational response been made, nor were tensions apparent. There are no discussions of the organisation of the engineering function in the light of CAD in the literature. However, the change seems to be mainly one of a gradual shift in the division of labour within a largely integrated part of the organisation(19). In particular, none of the separate developments of CAM and CAD threatened the respective functional and product organisation structures.

CAD/CAM and value engineering, on the other hand, integrate across this divide. They more closely couple the functionally organised manufacturing side with the product orientated engineering side. The first effect of this integration is to give the liaison role carried out by the production planning department a more crucial status. The department bacomes pivotal within the organisation and its influence spreads out in either direction. However, within the organisations studied the production planning department still retained its essentially functional orientation.

ORGANISATION DESIGN FOR CAD/CAM

A problem is clearly posed here - now that the two halves of the organisation are reciprocally coupled, their mutually incompatible organisation structures come into conflict. A number of writers have argued that it is possible to get the best of both worlds through implementing a 'matrix' organisation in which operating units report to both a functionally orientated and product orientated manager(20). Thus matrix organisations break with the traditional rule of a single line of command in order to integrate diverse elements of the organisation (see figure 3).

In cases B and D, signs of organisational stress were apparent in that the production planning functions had been reorganised more than once. In A, both halves were functionally organised, while in case C, a matrix type organisation structure had indeed been established (see figure 4). This company was by far the most developed in its use of CAD/CAM, both amongst these cases and nationally within the metalworking industries. While the relative simplicity of its product compared to the others and the particular energy of the innovators may well have been factors in establishing this lead, the integration of the organisation through the mechanism of a matrix organisation seems to have greatly aided the implementation of the new technology.

The phrase 'greatly aided' is used deliberately. The matrix organisation was established before the CAD/CAM system was installed, and had been under development since the mid sixties. The move was prompted by two problems. Firstly, there was a lack of co-ordination between sales and design on the engineering side; secondly, a lack of co-ordination between engineering and manufacturing. The response was

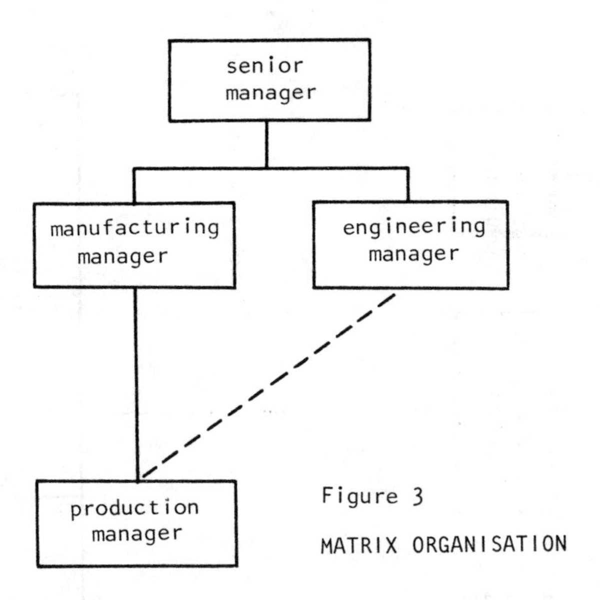

Figure 3

MATRIX ORGANISATION

to group engineering into product divisions containing sales, design and draughting, and accounts. This divisionalisation was resisted by manufacturing because of the risk to resource utilisation. Therefore, only the fitting shop was divisionalised, and placed under the responsibility of a product manager, who also acted as a progress chaser in the component shops. The production planning department also appointed personnel with particular responsibility for each product division. The product manager thus performs the key integrating role between the functionally orientated manufacturing side and the product orientated divisions within manufacturing. On the engineering side, integration has been achieved through the medium of 'product teams' which include the relevant product manager and planner from manufacturing. This last mechanism plays a crucial role in ensuring that new products can be manufactured efficiently. Design options are costed using data collated by the planning department and then evaluated in the light of this data.

MATRIX ORGANISATIONS IN ACTION

Matrix organisations can, therefore, handle the extra organisational linkages required for both CAD/CAM and value engineering.

Figure 4

MATRIX ORGANISATION: a simplified case - C

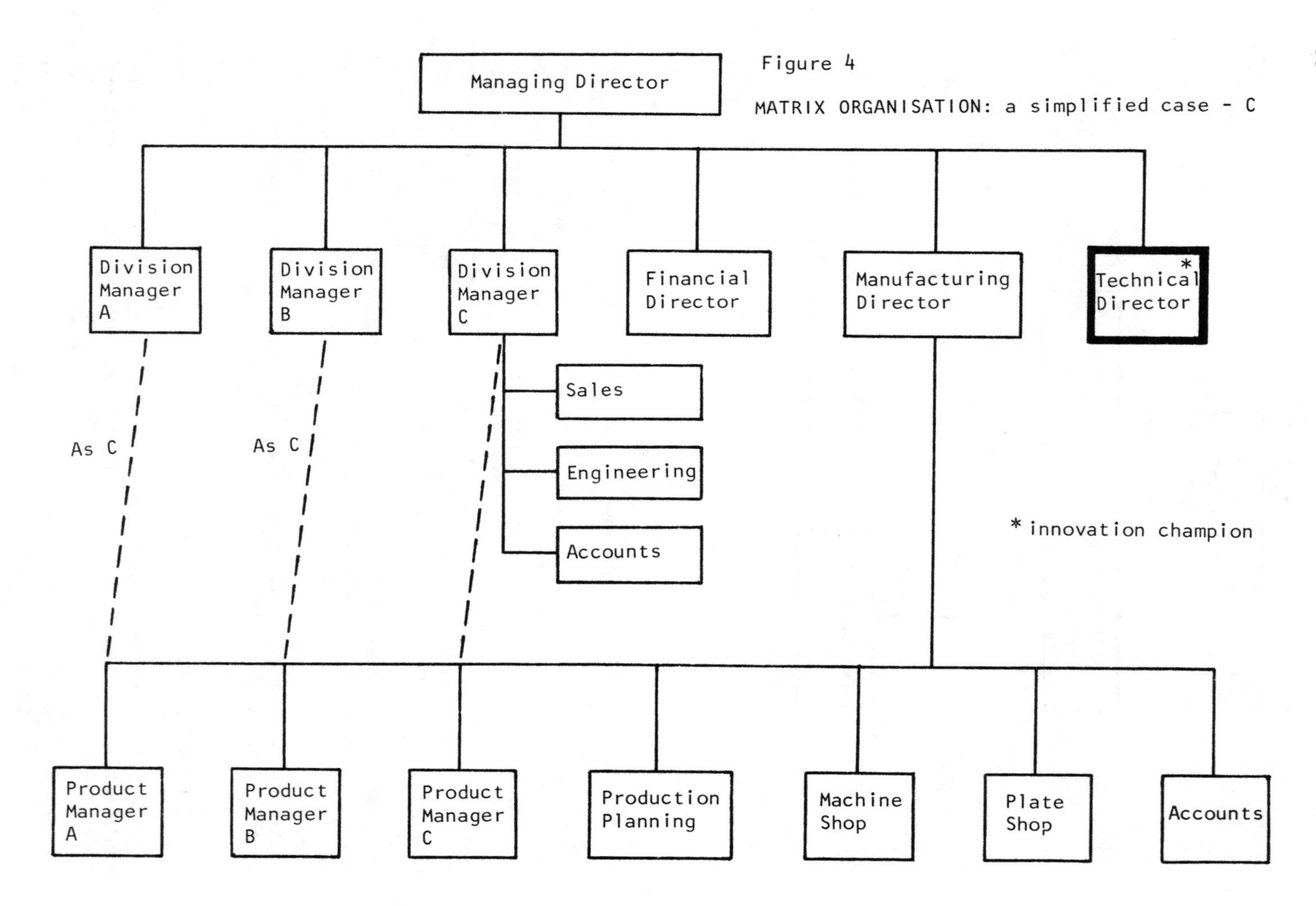

Galbraith in his study of Boeing(21) describes a number of organisational linkage mechanisms, many of which were developed as temporary solutions to short-term stresses such as running out of component stocks. More or less formal strategies such as liaison roles, task forces, and interdepartmental teams can all help develop organisational linkages, but CAD/CAM is a permanent development in the operations of a manufacturing organisation. This implies a permanent organisational solution. Matrix organisations are the most sophisticated way of institutionalising these organisational linkages implied by increasing reciprocal interdependence.

In the development of the matrix organisation, the production planning function is playing an ever more important role. It is pivotal in the information flows essential for the mutual adjustment in a reciprocal coupling. It is increasingly pre-specifying the work of the shop floor, with important implications for the role of the supervisor, skilled operator, and shop management. In case D, CNC programs were explicitly used by middle management to control the activities of junior management and operatives on the night shift. In the other direction, production planning is increasingly providing the data which are an important part of the appraisal for new designs from both the value engineering and NC manufacturing stand-points. It is likely that the power and influence of this department will increase through the eighties - the long march of Taylor's brainchild continues ever onward(22).

IMPLICATIONS

Two particular implications of this analysis are worth drawing out. The role of 'innovations champions' in stimulating the development and marketing of new products is well documented(23). The importance of such champions in process innovation was demonstrated by these cases. In cases A, B and C a particular manager played a crucial role in implementing CAD/CAM and NC. In case D, which is a larger operation than the others, a senior manager had responsibility for co-ordinating innovations in various parts of the organisation, but again proposals came from individual managers. In all cases the innovation champion was institutionalised in the sense that part or all of his brief was to develop process technology.

It was also clear that the organisational position of the champions is an important variable in their effectiveness. The positions of the innovation champions in cases B and C are shown in figures 2 and 4. Notably, in C, which had made the greatest progress with CAD/CAM, the champion was the technical director, and had a responsibility stretching right across both the engineering and manufacturing functions. In case B, and also in A, the champion was of a lower status and located within, respectively, the manufacturing and engineering functions. This meant that more progress had been made with CAM than CAD, or the other way round, depending on the champion's sphere of influence. In D, the senior co-ordinator relied upon proposals from champions in particular parts of the company which were then co-ordinated to ensure interface compatibility. Senker has argued that reliance on this sort of 'bottom up' innovation can distort the evaluation of CAD/CAM technology(24).

So far as the trade unions are concerned, this integration poses new problems. Workplace organisation has tended to follow the contours of the manufacturing process - with the white collar unions organising workers in information processing, and the blue collar unions organising those in materials processing. If the technology is bringing these two functions closer together, then it implies that there is greater need for joint action between staff and manual trade unions. This has already become an issue over NC programming, but the implications of this analysis are that the issues are more widespread. In none of the cases, however, had there been any liaison between the staff and manual unions over new technology at even an informal level, despite the reporting of relations in all cases to be good.

CONCLUSIONS

This analysis inevitably raises the old debate about technological determinism. What is being argued here is that matrix organisation complements CAD/CAM technology under certain conditions of market environment. If the environment is such that engineering can be functionally organised - as in one of these cases - then there is no need for organisational change. If, on the other hand, the technology is such that the economies of scale in manufacturing allow the abandonment of functional organisation, there is again no need for matrix organisation. In other words, it is the context of CAD/CAM production that leads to particular organisational implications. It is not argued that matrix organisations are only the result of CAD/CAM - the role of value engineering has been identified here, and often matrix organisation can be a solution to the problem of whether to organise a sales operation by product or geography(25).

However, the whole tenor of the argument implies that technology is considered an important organisational contingency. Woodward argued that it was in small batch and continuous process production that the impact of technology was strongest(26). The Aston studies, which stressed the importance of size as a contingency, did accept the influence of technology in those parts of the organisation closely associated with the workflow. In particular, they argued that technology was a big influence on 'workflow control'(27) - the stage three and four functions. More recently, Mintzberg has argued that so far as the main operating elements of the organisation are concerned, technology remains an important factor in organisation design(28).

It is probable that the pressures identified here will increase during the eighties. Tighter market conditions, due to the recession, are necessitating greater responsiveness to customers' requirements in the shape of shorter lead times, better quality, and lower cost. This implies product orientated organisation. On the other hand, the implementation of CAM, CNC and FMS on the manufacturing side is leading to greater capital intensity. This implies a functionally orientated organisation, and tighter control of resource allocation. Wilkinson has noted the way in which the implementation of a CAM system for production control led to a switch from product orientated progress chasing to functionally orientated shop scheduling(29). Matrix structure is likely to be an increasingly appropriate organisational form over the next few years.

So far as the debate about the impact of information technology is concerned, the implications of the above position are that firstly, it is important to locate the role of the new technologies in the existing production process - different elements of information technology do different things, and the context of their implementation is crucial for understanding the social and economic implications. Secondly, that the debate should be continued in the light of the existing body of sociological literature. Much of the early work on the impact of the chip has been extraordinarily naive and has shown little comprehension of the historical development of industrial relations, the labour process, or the strategy and structure of British industry. The tendency has been to extrapolate from the technologically defined potential of a piece of equipment, and to assume not only that this potential is inherently useful, but that its uses have far-reaching impacts throughout the industry. Technology is only one contingency amongst many others - in particular environment and size. It is the interplay of these variables, mediated through their perception by managers and trade unionists, which will lead to the shape of industry in the eighties.

Finally, the argument needs to be put in its wider context. The definition of efficiency implicitly underpinning the above argument is both _quantitative_ and _qualitative_. It is quantitative in the sense that matrix organisations provide the best means of maximising outputs from inputs in batch metalworking in certain capitalist market environments. It is qualitative in the sense that the organisational solution under capitalism re-inforces the hierarchical character of the firm. The _co-ordination_ achieved by the growing role of the production planning function is also _control_ of the labour process in information and materials processing.

AUTHOR'S NOTE

The research reported here was conducted in collaboration with Arthur Francis, Mandy Snell, and Paul Willman. I would also like to thank Craig Littler for encouraging me to write this chapter. The research was financed by the Joint Committee of the SSRC and SERC.

REFERENCES

1. See ADVISORY COUNCIL FOR APPLIED RESEARCH AND DEVELOPMENT. Computer Aided Design and Manufacture. London: HMSO, 1980; BESANT, C.B. Computer Aided Design and Manufacture. Chichester: Ellis Horwood, 1980; and PRESSMAN, R.S. and WILLIAMS, J.E. Numerical Control and Computer Aided Manufacturing. New York: John Wiley, 1977.

2. See HATVANY, J. The Distribution of Functions in Manufacturing Systems. in: McPHERSON, D. (ed). Advances in Computer Aided Manufacturing. Amsterdam: North-Holland, 1977; and ABELL, P. and MATHEW, D. The Task Analysis Framework in Organizational Analysis. in: WARNER, M. (ed). The Sociology of the Workplace. London: Allen and Unwin, 1973.

3. See KYNASTON REEVES, T. and TURNER, B.A. A Theory of Organization and Behavior in Batch Production Factories. Administrative Science Quarterly, 19, 1972

4. GORDON, D.M. Capitalist Efficiency and Socialist Efficiency. Monthly Review, 28, 1976.

5. See MINTZBERG, H. The Structuring of Organizations. Englewood Cliffs: Prentice-Hall, 1979, Chapter 7; and the discussion of 'Printer Inc' in LORSCH, J.W. and LAWRENCE, P.R. Organization Planning: Cases and Concepts. Homewood: Irwin-Dorsey, 1972

6. THOMPSON, J.D. Organizations in Action. New York: McGraw-Hill, 1967, Chapter 5.

7. KLEIN, L. Multiproducts Ltd. London: HMSO, 1964.

8. KYNASTON REEVES and TURNER, op cit.

9. ADVISORY COUNCIL FOR APPLIED RESEARCH AND DEVELOPMENT, op cit, S.2.14.

10. THOMPSON, op cit, Chapter 5.

11. WOODWARD, J. Industrial Organization: Theory and Practice. London: Oxford University Press, 1965, Chapter 8; and KYNASTON REEVES and TURNER, op cit, p.93.

12. Cited COLCHESTER, N. Productivity Cliches Fail the Test. Financial Times, 20/12/77.

13. See LOCK, D. Engineer's Handbook of Management Techniques. Epping: Gower, 1973.

14. WILLIAMS, L.K. and WILLIAMS, E.B. The Impact of Numerically Controlled Equipment on Factory Organisation. California Management Review, 7, 1964.

15. BRAVERMAN, H. Labor and Monopoly Capital. New York: Monthly Review Press, 1974; see also WOOD, S. (ed). The Degradation of Work? London: Hutchinson, 1982.

16. See LITTLER, C.The Development of the Labour Process in Capitalist Societies. London: Heinemann, 1982; and CHILD,J. and PARTRIDGE, B. Lost Managers. London: Cambridge University Press, 1982.

17. COOLEY, M. Computer Aided Design - Its Nature and Implications. London: AUEW (TASS), 1972.

18. COOLEY, ibid, p.77.

19. See SENKER, P. and ARNOLD, E. Designing the Future. Watford: EITB, 1982.

20. See GALBRAITH, J.R. Organization Design. Reading: Addison Wesley, 1977, chapter 10; and MINTZBERG, op cit, chapter 10.

21. GALBRAITH, J.R. Environmental and Technological Determinants of Organisational Design. in: J.W. LORSCH and P.R. LAWRENCE (eds). Studies in Organisation Design. Homewood: Irwin-Dorsey, 1970.

22. For an analysis of the subtleties of these developments in the stage 3 process in shipbuilding - 'lofting' - see BUCHANAN, D.A., and BODDY, D. Organizations in the Computer Age. Aldershot: Gower 1983, chapter 8.

23. LANGRISH, J., GIBBONS, M., EVANS,W.G., and JEVONS,F.R. Wealth from Knowledge. London: Macmillan, 1972.

24. SENKER, P. Some Problems in Justifying CAD/CAM. Paper presented at Automan '83, Birmingham, May 1983.

25. See MINTZBERG, op cit, p.171.

26. WOODWARD, op cit.

27. HICKSON, D.J., PUGH, D.S. and PHEYSEY, D.C. Operations Technology and Organizational Structure: An Empirical Reappraisal. Administrative Science Quarterly, 14, 1969; and CHILD, J. and MANSFIELD, R. Technology, Size and Organization Structure. Sociology, 6, 1972.

28. MINTZBERG, op cit, chapter 14.

29. WILKINSON, B. The Shop Floor Politics of New Technology. London: Heinemann, 1983, chapter 8.

Technological imperatives and strategic choice

David Buchanan

INTRODUCTION

This paper is based on case studies of the impact of five different types of computing and information technologies in four manufacturing companies in central Scotland:

Company:	New technology:
Caterpillar Tractor Company	numerically controlled machine tools computer coordinate measuring machines
Govan Shipbuilders	computer aided lofting
United Biscuits	computerised equipment controls
Ciba-Geigy	computer aided process controls

These studies were concerned with process rather than product innovations and the researchers' main interests were in the human, organisational and managerial consequences of technical change.

The label 'information technology' is now attached to a wide range of different types of machinery. The publicity surrounding these technical developments gives a false impression of the monolithic nature of their capabilities. Microprocessors and other computer components are simply building bricks from which a diversity of devices can be constructed. It is important to note therefore that the conclusions of this research are based on studies of five particular manufacturing systems and the results may not be fully generalisable to other manifestations of information technology.

One choice in presenting such research results is to deal either with the details of individual cases or with the main line of argument.

When the former mode is adopted, it is easy to be quickly lost in the mysteries of lofting in shipbuilding or in the capabilities of coordinate measuring machines, and thus lose sight of the general conclusions. This paper concentrates on the main argument, and consequently loses sight of the detailed empirical base. The detail may however be found in another publication(1).

THE MAIN ARGUMENT

The main conclusion from this research is that technical change acts as a trigger to processes of management decision making. The choices that form in those decision making processes, concerning why and how the technology is to be used, determine the outcomes of technical change. Technology has no impact on people or performance in an organisation independent of the purposes of those who would use it and the responses of those who have to operate it. The technologial imperatives are weak; the strategic choices are crucial. This argument has been stated by Child(2) and by Clegg and Dunkerly(3), and is elaborated in Figure 1.

Technological determinism is therefore not a useful perspective. It does, however pervade much of the commentary concerning contemporary information and computing technologies. The advantages of these innovations are not automatic and are not necessarily rapidly achieved. Media coverage and British government activities in the early 1980s may have generated unnecessary fears and unrealistic expectations. These false assumptions may influence the effectiveness of future applications.

The process or technical change in the companies studied had some noteworthy features. None of these companies had entered into formal consultations or negotiations with their workforce with respect to the technical changes introduced. One company has subsequently entered into negotiations over a 'new technology' agreement with their unions concerning other technical changes. None of these companies faced organised resistance to change from their workforces. In one case, management resisted the introduction of new machinery. None of these companies made any employees redundant as a direct result of these technical changes.

In each case it was possible to identify clearly a 'promoter' or 'champion' who persuaded others of the value of the change. In every case there was differential support from different levels and functions within the organisation; individuals' perceptions and expectations of the new technology were coloured by their experience and position within the organisation. In most cases the strongest reasons behind the investment in new machinery lay with factors internal to the organisation, rather than with external factors such as competition or changing consumer tastes.

WHAT ? THE CAPABILITIES OF INFORMATION TECHNOLOGY

These technologies have four information handling capabilities:

-information capture:	They gather, collect, monitor, detect and measure.
-information storage:	They convert numerical and textual information to binary, digital form and retain it in some form of memory from which information can be retrieved when needed.
-information manipulation:	They can rearrange and perform calculations on stored information.
-information distribution:	They can transmit and display information electronically, on screens and on paper.

But as the purpose of the information handling is usually to effect control in some respect, 'control technology' is perhaps a more appropriate label.

Braverman(4) argues that '...the key element in the evolution of machinery is not its size, complexity, or speed of operation, but the manner in which its operations are controlled'. Computing technologies do this in three broad ways

-the equipment gives the operator feedback information to make operator control of the equipment or process more effective.

-the equipment or process is taken under computer control through a predetermined sequence or cycle of operations.

-deviations of equipment from process standards are measured and corrective action initiated by the computer.

All three of these features have been termed 'automation', a word that has created much misunderstanding. Bright's(5) definition of automation is more useful, meaning simply, 'more automatic than previously existed'. Many so called 'automatic' systems are in practice dependent on skilled and knowledgeable human intervention for their effective operation.

These capabilities and features are enabling characteristics. They open up new opportunities, possibilities and areas of choice for products, processes, and organisational arrangements. The ability to provide operators with performance feedback information is particularly important in this respect. These capabilities do not on their own "determine" organisational functions and structures.

WHY ? MANAGEMENT OBJECTIVES

The managers who introduced these technologies were interviewed to find out why and how they had done so. The opinions recorded here reflect the thinking of the managers on the research sites at the time of the study and do not necessarily reflect the current policies of the companies concerned.

When asked why, managers spoke of three types of objectives for introducing new technology.

1) Strategic objectives:

Several external, economic, market and customer oriented objectives appeared to be important, such as the desire to:

- upgrade machining and inspection equipment to meet changing market conditions;
- reduce preproduction lead times;
- attract customers by being market leaders in the use of new technology;
- carry out company policy to maintain investment in new technology to reduce costs and improve product quality;
- increase capacity to meet market demand;
- improve product consistency and price compared with competition.

2) Operating objectives

Several internal, technical, performance oriented objectives were also important, such as the desires to:

- reduce production costs;
- replace obsolescent equipment;
- overcome bottlenecks in production created by other changes;
- computerise one part of a process to exploit fully other related technical changes;
- control energy use and cut other plant running costs;
- improve the flexibility of the production process;
- reduce numbers and costs of support staffs;
- reduce product waste.

These operating objectives are clearly related to and consistent with strategic objectives.

3) Control objectives:

These objectives include the desire to:

- reduce human intervention;
- replace people with machines;
- reduce dependence on human control of equipment and processes;
- reduce uncertainty, increase reliability, predictability, consistency and order in production operations;

- increase the amount of performance information and the speed at which it is generated.

Strategic thinking was characteristic of senior management. Operating concerns were characteristic of middle line and financial management. The main concern of middle and junior line management was control.

HOW ? THE ORGANISATION OF WORK

The task of organising work around new technology is usually the concern of middle and junior line management. Their pursuit of control objectives in some cases led to the design of forms of work organisation that gave operators neither the skills not the motivation to perform their functions effectively. Control objectives are therefore not necessarily consistent with strategic and operating objectives.

Technology can either replace human effort and skill, or it can complement it. The household washing machine and the industrial robot are examples of replacement. Musical instruments and conventional machine tools are examples of complementarity.

In biscuit making, the traditional craft skills of mixing biscuit dough were replaced by a computer that stored biscuit recipes and automatically fed correct amounts of the appropriate ingredients to enclosed mixing vessels. Management changed the job title from doughman to mixer operator.

The mixer operator was required to start the mixing machine on the instruction of the computer, and to empty the mix into a hopper twenty minutes later when the mix was over. The mix was interrupted two or three times to allow the operator to add small quantities of other ingredients such as salt and other chemicals which had proved difficult to insert automatically. Thus although the doughman's skill had been replaced, the need for human intervention was not removed. Previously, the doughman could see, hear, and feel the dough as it mixed and could when necessary add ingredients to adjust the quality. This was no longer possible. The mixer operators:

- had a limited understanding of the process and equipment;
- could not visualise the consequences of their actions for other stages of the process;
- could not trace sources or diagnose causes of equipment faults;
- developed no knowledge or skill that could make them promotable.

So although the dough output was more consistent:

- specialist maintenance staff were required;
- the mixer operators became bored, apathetic and careless;
- the mixer operators rejected responsibility for breakdowns;
- the management lost a source of supervisory recruitment.

This key, but 'residual' or 'distanced' role of the human operator may be characteristic of jobs in 'nearly automated production systems'. The step from this nearly automated stage to one in which human intervention is completely absent is not a small, incremental one but may involve significant advances in production technology. The completely unmanned factory may be a rare or special case.

The organisation of the work of the mixer operator reflected management control objectives. But a different physical layout and a reorganisation of work on the production line could have complemented the skills of the doughman, avoided the distancing effect, and still retained the advantages of the new computer system.

The mixer operator could have been given responsibility for the computer controls. At the time of the study, the conventional computer system was being replaced by a small, easy to operate, microprocessor controlled 'recipe desk' through which biscuit recipe changes could be made by adjustments to small thumbwheels. Operators could have been organised into 'line teams' (which would include a mixer operator, a dough cutting machine operator, an ovensman and a biscuit packing group), each with an overall responsibility for their process and product.

But this would have been difficult to implement for several reasons. The physical plant layout and organisation structures created when the computer system was first installed inhibited the adoption of an alternative form of work organisation. The recipe desks went into the computer operators' room along with the previous computer which was being replaced gradually. The computer and line operators belonged to different trade unions. The payment system differentiated operators at difficult positions on each production line. Management decision making may thus be constrained by the results of past decisions as well as by current objectives.

Information technologies can complement human skills, particularly those concerning information processing, problem solving and decision making. Computers are used extensively in chemicals manufacturing to monitor and control process variances. But characteristics of the product and the process equipment inhibit the removal of human intervention.

In the chemicals plant studied, unsupervised operators made process control decisions on the basis of their experience, judgement and intuition, in response to computer generated information that was uncomplete and sometimes incorrect. The plant operators were required to interpret the pattern of information available to them at any given time, and to override or adjust the computer controls when appropriate. If process control were left entirely to the computer system, the frequency of plant shutdowns would have been intolerably high. It was not possible to put computer sensors on every part of every piece of equipment that might go wrong. Sensors became fouled with the messy product and gave false alarm readings. The minimum training time for these skilled plant operators was one year. The productivity of the plant studied was twice that of comparable conventional plants at the same site.

Research has demonstrated(6) that the organisation of work is effective in terms of motivation and productivity, where the knowledge and skills of operators are complemented by providing:

- information feedback on performance;
- meaningful, interesting and challenging work;
- control over workflow;
- discretion over methods and task allocations;
- opportunities to develop skills and knowledge.

The information handling capabilities of computing technologies create new opportunities and possibilities for the reorganisation of work in ways that are effective on technical, economic, and human criteria. Management may prevent this by concentrating on control objectives which may be unrealistic given the current state of technology.

CONSEQUENCES AND CONCLUSIONS

This research suggests that information technologies can be used to achieve a wide range of organisational objectives. But the strategic gains may be overlooked in the pursuit of operating and management control objectives. In one case, managers' resistance to the introduction of new technology was based on the belief that they would lose control over the workflow through the sections for which they were responsible. The impact of information technology may thus depend to a large extent on how managers relate its application to broader business objectives and to the roles and functions of management itself.

This research also suggests that human and organisational implications are often disregarded or their importance underestimated, although they have a significant effect on the degree of success of technical change. The organisational choices that accompany applications of new technology are rarely recognised and evaluated. Physical layouts and organisational structures created by past decisions tend to be taken as given and the possibilities and opportunities of new arrangements are not explored.

Technical changes are usually organisational changes. Those directly affected by a particular application typically alter their work rates and procedures in ways that indirectly affect other occupational groups. The disciplines introduced with computer based systems heighten the importance of interdependencies between sections of an organisation. The efforts of technical change are thus rarely confined to one group or department. The research evidence also suggests that the implications or these 'knock on' or 'ripple' effects are rarely considered.

Technical change would perhaps generate fewer frustrations and disappointments if human, organisational, layout and ripple factors could be taken into account in management decisions about why and how to apply new technology.

Organisations are political systems in which people compete with each other for resources such as status, influence and money. Different 'stakeholders' have different and conflicting perceptions and interests. Expectations of, and reactions to, any kind of change reflect the position and experience of different individuals and groups in an organisation. Since information is a resource which confers power, status and influence on the possessor, applications of information technology may heighten the inherent conflicts of interest between stakeholder groups.

Information technologies alter patterns of access to information in organisations. Their introduction may therefore threaten the traditional boundaries between operating and managing roles. The operator of computerised equipment is difficult to supervise because the operations that matter happen silently and invisibly inside the machinery. In some instances in this research, management had become dependent on the knowledge and expertise of equipment operators and on technical support groups which maintained and programmed the new technology. In some instances, operators had access to the same performance information as management - information captured, analysed and distributed automatically by the computing technology.

The advantages of new technology may be lost if it is applied with traditional management values and assumptions. The effective use of new information and computing technologies may be dependent on new forms of work organisation. This may in turn necessitate a reconstruction of the role of management, particularly at lower levels of line management.

But this need not be a game in which some can win only if others lose. The advantages of technical change may be achieved and the problems avoided by careful consideration of the full implications of change. Future research should perhaps explore ways of analysing and presenting the technical and organisational issues and choices in ways that can help stakeholders to explore mutually advantageous ways of exploiting the enabling characteristics of information technology.

REFERENCES

1. BUCHANAN, D.A. and BODDY, D. Organizations in the Computer Age: Technological Imperatives and Strategic Choice. Aldershot: Gower, 1983.

2. CHILD, J. Organization Structure, Environment and Performance: the Role of Strategic Choice. Sociology 6(1), 1972, 1-22.

3. CLEGG, S. and DUNKERLEY, D. Organization, Class and Control. London: Routledge and Kegan Paul, 1980.

4. BRAVERMAN, H. 1974, Labor and Monopoly Capital: the Degradation of Work in the Twentieth Century, New York: Monthly Review Press, 1974.

5. BRIGHT, J. Automation and Management, Boston: Harvard University Press , 1958.

6. BUCHANAN, D. The Development of Job Design Theories and Techniques. Farnborough: Saxon House, 1979.

Computerised machine tools, manpower training and skill polarisation: a study of British and West German manufacturing firms

Ian Nicholas, Malcolm Warner, Arndt Sorge, and Gert Hartmann

RATIONALE

Since the fifties metal-cutting has undergone major technical advances. It has, in the eyes of the laity, been 'automated'. The development of Numerical Control (NC) of machine tools, that is, the type of machine where pre-determined machining sequences and cutting operations are 'fed' into the machine by some form of previously-prepared programming tape, was followed gradually by the introduction of Computer Numerical Control (CNC) equipment where the machining operations are controlled by the use of 'online' computers and other microprocessing technology. An alleged consequence of these developments has been the polarisation of skills with increasing 'deskilling' being postulated for the shopfloor workers(1). But has technological change, and as importantly, its organisational and manpower implications, in fact, taken such a dramatic path?

Innovation may produce new dimensions in the variables very gradually, but this does not reduce its importance: 'the devil is in the details' as an old German phrase puts it. The theoretical rationale for this view is the proposition that technology affects organisation and manpower in an interactive pattern rather than any direct causal way(2). The elements of the national culture, with its implicit work traditions, shaped by education and training values and institutions, amongst other factors, in turn constitute the parameters of this pattern.

While our study did not examine trends over time quantitatively, such assessments were implicit in the interview data. The trend to using more skilled labour may be anticipated in the future, and more strongly in Germany, because of a relatively greater supply of trained manpower present at all levels(3). In Germany, for example, over half the primary workforce has gone through a formalised apprenticeship, and this institutionalised qualification system in Germany, as well as the works council system which it reflects, is likely to push the process further.

We were, however, aware that industrial relations differences between Britain and Germany did not appear to be central to greater or lesser success of NC adoption(4), and in Britain itself, NC and CNC use did not, for example, typically lead to demarcation disputes(5). Hence, we did not stress collection of industrial relations data as such. Typical industrial relations differences, as described in detail by Marsh et al (6), are in reality closely linked with variations in manpower training and organisation(7), but they cannot be conjectured as having a strong independent influence on differences in NC and CNC utilisation. It is with these broad conjectures in mind, building on our earlier work, and anticipating a cumulative contribution to knowledge in the field, that we now turn to the modus operandi of the empirical study carried out.

THE METHOD OF PAIRED COMPARISONS

Our work was influenced by the methods of, in particular, the Franco-German comparison of organisation and manpower developed by Maurice, Sellier and Silvestre(8), and the partial replication for Britain, France and West Germany(9). These employed a range of various criteria for matching such as size, product, technology of production, dependence, and urbanisation. In the present study, because of detailed concern with the division of labour of the shop-floor, the matching process was directed most carefully towards product and production techniques. 'Perfect' matching, however, with regard to production technology, was excluded because our study was explicitly designed to describe and explain the piecemeal evolution of technology and one of the major objectives was to highlight the different stages of evolution of CNC technology across the various organisations.

Some deviations from absolute matching standards were tolerated when we were convinced, beyond reasonable doubt from exploratory interviews, expert evidence and initial visits to sites, that any deviation was not liable to influence organisational and labour differences. On the other hand, if there was such a suspicion, a unit was not included. Whilst this led to a rather lengthy search for matching units in the two countries, it ultimately proved helpful in gaining greater familiarity with the general terrain and in helping to generate useful hypotheses.

In turn, this knowledge about the wide-ranging diversity of CNC techniques, the increasing range of innovation across the batch size and components spectrum, and the widening range of use, was used to define a series of cases of application which could be compared both within countries and between countries. Average batch size (or batch running time) and plant size were hypothesised as influencing organisation and manpower variables; similarly, other more qualitative measures were used. These included component complexity, type of machine, degree of precision, scheduling procedures, programming arrangements and so on. It needs to be emphasised, however, that these technical differences, rather than being used simply as matching criteria, needed to be explained. They were seen as different dimensions arising from a socio-technical interaction between technology, organisation and labour.

To this end, the data collection process was wide ranging. It followed an agreed series of guidelines rather than a ready-made questionnaire because of the necessity to adapt the questions to the circumstances, terminologies, structures and procedures prevailing within each unit. In 'expert' interviews we asked questions about:

- product and process development
- organisational philosophy and the role of different departments
- perception of rationale of CNC use in general
- justification of CNC acquisition and use in the company, in economic terms
- maintenance, after-sales service, and training provided by manufacturers
- personnel policy and industrial relations
- appropriate training and qualification
- problems of the labour market in general and that affecting the company

In employee interviews, we concentrated on:

- work tasks of the interviewee
- collaboration with people of different jobs and departments
- characteristics of work with CNC equipment as compared to other machines
- perception of training necessities and provision
- experience in the present job as compared to previous jobs

RESEARCH FINDINGS

General

Was technology more of an intermediate factor than a crudely determining one(10), at least in the matched case-studies examined?

It was striking that the predominance of the technology examined varied with the size of plants studied. Both between small and large plants, and within plants, it was moreover difficult to pin-point a particular phase of NC/CNC development because the technology evolved gradually and, in many cases, the gamut of technical evolution was visible in the factory site in question.

Before showing the way in which we found the polarisation of skills not to be the inevitable consequence of the new technology, we must look first at the production techniques themselves and also at the organisation structures we found. After this, the effects on training and employee qualifications will be discussed, as well as on personnel structures, in order to ascertain the way the polarisation of skills can be characterised.

Production Techniques

The small plants studied all had a high proportion of NC and CNC machinery, approaching, in some cases, 100% of machine capacity. The larger plants show a predominance of other machinery, with on occasions the absolute numbers of CNC machnies being smaller than those of small plants.

The larger plants in Germany had more CNC machines than in Britain, but this difference was frequently counteracted by the presence of additional NC machines. Large plants in both countries typically had a longer and more continuous experience with a range of NC machines, whereas the small plants usually went straight to comparitively-recent CNC equipment, purchased in substantial numbers. This is particularly significant in view of the substantial capital outlay involved compared with the resources available to the small company. In general, however, the 'traditional' innovators, particularly for CNC, in both countries, at least in the firms studied, were small companies.

The more prolific application of CNC in the German larger companies was mainly due to the use of CNC lathes. The widespread use of NC in turning was a more recent phenomenon whereas the earlier focus of NC had been on milling applications. Such CNC lathes however consistently featured readily adaptable and easy-to-use shop-floor programming facilities.

Thus, the diffusion of CNC, in our selection of companies, appeared to depend on:

- small size of a company
- emphasis of CNC turning vis-a-vis milling
- shop-floor programming facilities

The justification of CNC ex ante was in rather general terms often intuitive through lack of information about manning requirements, down-time, maintenance and service quality, equipment reliability, organisational implications and so on. Ex ante and ex post, CNC use in general was justified on the basis of the complexity of the geometric design of the components and the required cutting cycles and sequences. Economic justifications were also discussed by respondents but were often presented as secondary considerations. Arguments for CNC, however, also emphasised increased flexibility coupled with potential increases in productivity in manufacturing. In the past, productivity increases had involved the production process becoming less flexible and less capable of handling variations in component specifications, small batch size, batch conversion and so on.

Differences were seen to exist in the character of the production technologies selected in the German and British plants. It was, for example, rather more difficult in the case of both small and bigger plants to find examples of large batch processing involving CNC equipment in England. This may have been because of the recession reducing batch sizes. We do not interpret this wholly as a problem of matching: it would seem that there are different nationally specific overall socio-economic trends in operation which affect the application of CNC both quantitatively (more CNC turning in Germany) and qualitatively.

Production process

Again, vis-a-vis skill de-polarisation in both countries, operators retained the setting and programming-related functions to a greater extent when batch size was small. Larger batches inevitably meant that operating was more differentiated from the preparatory functions which, in these instances, were expanded and generally referred to planning and scheduling. Larger plants were more likely to have separate planning and programming departments.

In small plants, setting and programming-related functions were intimately combined. In the small plant/small batch classification, there was a large overlap of operating, setting and programming-related functions. In the small plant/large batch category, setting and programming were still closely combined but they were differentiated from the normal running of a batch. The policy, however, in both countries, with some differences, was to encourage - and this is important in the context of the de-skilling debate - the development of operator expertise in the programming-relating functions. This was not so much an end in itself but it had been recognised that the operator was uniquely placed, via the CNC 'electronics', to control and, if necessary, modify the cutting process.

Operators said they were more concerned with the problems of metal removal, that is, feeds and speeds, tool selection, tip wear and quality, tool life and so on, than with programming difficulties. Some limitations with machine-programming were nevertheless encountered; these were usually associated with the question of reduced machine utilisation that might result from the direct programming of geometrically difficult components.

Within all the cases, British companies used NC and CNC in such a way as either to maintain planning department control and/or autonomy or to segregate the operations from the other sections. In Germany, NC and CNC organisation was fashioned so that it linked foremen, chargehands, workers, and planners around a common concern. This was particularly apparent in the case of foremen, which in Germany were deeply involved in CNC expertise, but who in England had been largely by-passed, resulting in de-skilling of the job.

In both countries, the attractiveness of programming-related functions lay less in a formalised language, but in the instrumental value of programming as a means whereby the more demanding part of the task, that of metal cutting, could be more effectively planned and controlled. This diffusion of control to the shop-floor depended on the ability to make programs on the machines without losing too much machining time, but more importantly on the accepted view of operator competence in metal-removal technology. This view was more partial to physically locating program-related functions in the hands of planners in Britain whereas it was more likely to prefer programming by operators as experts in cutting and setting in Germany.

Organisation

In the smaller plants studied in both countries, organisational structures were decidedly informal, which was hardly surprising. Production engineering and management expertise were linked in the person of a visible owner or director. In one pair of smaller plants, some organisational differences between countries were apparent; in the British plant there was a differentiation between production management and production engineering, and similarly there was also a distinction between CNC and conventional operations.

Larger plants were more likely to have organisation charts or some other formalised structures. Formal organisation charts tended, however, to be less employed in German companies, at least as far as internal use was concerned, where they are seen as bringing about unnecessary dissension over status. The difference between Germany and Britain regarding the production/production engineering link or division, more visibly emerged here, particularly in the small batch/large plant category. Programming was placed within the work planning department and was closely linked to the production control function: in Germany all of these groups were more integrated into the line of production management whilst in Britain they tended to be structurally isolated, or at least, rather more remote. In one case in Germany, following the large-scale introduction of CNC, programming was relocated, both organisationally and physically, to be closer to the shop-floor and subordinate to a lower level of production management.

In the large batch/large plant classification, organisational structure was further differentiated. Nevertheless, the German company used production engineering functions as a link between differentiated production, work planning and production control departments. The British company, with a very complex organisational structure, tended more to view the programming, planning and production manufacturing and industrial engineering groups as discrete 'support' activities.

The greater degree of emphasis on the 'centrality' of engineering and production in German plants is also evidenced in other ways. Managerial control, for example, over different divisions or production sites is more intensive in technical respects, that is, in terms of product and process development and the achievement of production targets, deadlines and delivery dates. In Britain, by contrast, control occurs primarily through financial performance targets and indicators where the emphasis is on financial return. These differences, however, are not simply attitudinal; they extend to differences within the financial and economic institutions within the two countries. German companies, to a greater extent, tend to finance investment in new equipment out of bank loans, whereas in Britain, companies appear, for this purpose, to rely more on either the generation of 'internal' funds or on other forms of equity capital. This comparison also extends to the question of the economic justification of new equipment: it has often been observed that British companies expect new plant to pay for itself more quickly than similar firms in Germany. This reinforces the view that British companies tend to emphasise profits as a company goal whereas German firms stress the technical and production preconditions necessary to achieve returns.

Training

Rather than any one deterministic outcome, there was a striking variety in the training and qualification patterns under CNC application, which was closely related to differences in the organisation of the work process. There were no unambiguous qualificational consequences arising from the use of CNC, and we can go further and observe that CNC offered the possibility of less polarised qualification structures than under NC, noting that it had been developed partly with the idea of making de-polarisation possible. De-polarisation is, nevertheless, not a necessary consequence, but indeed hinges on socio-economic and institutional factors, some of which are nationally specific.

In both countries, using more skilled people as CNC operators was strongly linked to the integration of the operating and machine-setting activities. Where this existed, operators were more likely to be skilled; where it was absent, operators were unskilled. This relationship, however, was also a function of batch size, and thus it followed that there was also a differentiation between semi-skilled operators and skilled setters where large batches were run, but more skilled operators, who also performed setting functions, prevailed in small batch processing. In general, the greater the batch size, the greater the qualificational polarisation of the workforce.

The most pervasive national differences occurred in the small batch/large plant category. German companies allied CNC operations to tradesman status and experience, but British companies did not necessarily follow this pattern. Where this was so, however, it was usually recently introduced and not uncontested within the company. There was a consistently held view in Britain that CNC tended towards de-skilling and that CNC operation was more routine. The same companies are however planning to introduce skilled tradesmen onto their NC and CNC machines in an attempt to increase machine utilisation and, to a lesser extent, product quality.

Personnel

Personnel structure, in terms of white-collar staff and blue-collar worker ratios, under CNC application, was the result of a complex set of influences which did not necessarily originate from CNC. There was a general tendency in most of the companies investigated, particularly the small British but including most of the German companies, for there to have been an assimilation of status between the shop-floor and the planning functions. The personnel structure remained more or less constant through the action of the status-bound grading policy, and any shift in the distribution of actual functions had limited influence. This status assimilation may have even helped advance the utilisation of the CNC process.

However, this was not the complete story; assimilation took on very different forms. When craft qualifications were more frequently held on the shop-floor, and particularly on the part of the operators, the stronger was the tendency to class programmers as workers or to integrate programming and planning functions into the tasks of these workers. Conversely, the weaker the craft qualifications, the stronger was the tendency to achieve this assimilation by reclassifying shop-floor personnel as white-collar staff. Both assimilation effects could be observed in Germany; but their respective incidence depended on batch size which, in turn, was strongly related to operator qualifications. In short batches, operators progressed onto programming tasks without changing status, whilst with large batches, white-collar status was extended into the works.

Assimilation of status in Britain took place in small companies, but was less so in larger factories where there was a more consistent distinction between works and staff. This was despite the more legalistic character of the definition of workers and white-collar employees in Germany.

Where figures are comparable, Britain used craft workers more sparingly than Germany. These differences in personnel structure were related to the degree to which production engineering, work planning, and production control tasks were set away from the shop-floor.

The relationship between white-collar employees in programming and production control, and the shop-floor, changed very little even with intensive CNC use. The demand however shifted towards flexibility and productivity increases as the result of shop-floor activities which may be seen as a function of production expertise. In Germany this was more visible on the shop-floor and the associated functions moved to meet it on the 'home' ground. In Britain, the programming, planning and control staff were seen as the 'proprietors' of such knowledge, but this philosophy was being questioned more and more critically, as in Germany, if not for the same reasons.

SUMMING UP

Commonalities and differences

There is, thus, a series of very clear 'logics' of CNC application. These different concepts are analytically distinct, but they interact, and sometimes conflict, with each other. To distinguish between the available alternatives, we need to look at:

- Company or plant size;
- Batch size, or conversely, the time needed to machine a batch;
- Complexity of component, metal removal technology, and machine type;
- National institutions and accepted practices of management and in the

nature of technical work, training, status differentials, etc.
- Socio-economic conditions of the present situation, regarding shortage of natural resources, limitation of mass markets, slow growth, market competition, etc.

Company size

'High technology' by no means signifies greater bureaucratisation; the smaller plants studied combined a strikingly higher percentage of CNC machinery with personalised industrial relations, weak formal methods or organisation, and traditional entrepreneurial, paternalistic style, whereas the bigger plants were sometimes prevented from larger scale CNC use by conventional organisation and industrial relations of a more bureaucratic type. Whilst the bigger plants were starting to incorporate CNC into apprenticeships and other training schedules in a more systematic fashion, this was rarer in the smaller plants. Small companies profited from the qualifications and expertise inherent to personnel who were trained or who worked in larger plants, and although this knowledge was not developed formally, they were ingenious in pushing ahead without organisational shackles.

Batch size

Whilst plant size was associated with a greater differentiation of programming function into separate departments, increasing batch size was linked with the greater differentiation of programming away from machine-operating, although not necessarily into separate departments(11). Nor was it necessarily taken away from the machine; there was a strong inclination for program input on the machine, not by the operator, but by a programmer, foreman, or setter, in small companies producing large batches. This was the case of an unbureaucratic, but polarizing division of labour and qualification structure. It is important to distinguish between bureaucratisation within the organisation, and the polarisation of skills; increasing plant size is linked with the former, while increasing batch size is related to the latter. They interact, however, to bring forth organisational and qualification patterns (see table 1).

This scheme traces possible solutions, around which various modifications occur. It can be readily, and intuitively grasped that the smaller the batch size, and greater is the need for frequent conversion of machines with new tools, fixtures and component programmes, and the less is the machine-setting expertise differentiated from operating. Batch size economies of scale thus involve a reduction of skills at the operator level, a factor which has no doubt coloured the 'pessimistic' literature in the field *vis-a-vis* deskilling.

Machine-setting and program-related functions, however, overlap; the elimination of setting from machine operation also entails differentiation of program-related functions from operating, and it is often linked with a further differentiation between setting and programming.

Table 1

Skills, polarisation and bureaucratisation of programming under CNC

		Plant size: bureaucratisation of programming	
		small	large
Batch size polarisation of functions and skills	small	programming with larger involvement of operators, and small or non-existent programming department	programming more in the hands of a programming/planning department, but skills of operators and planners overlap
	large	programming done by setters, foremen, or programmer; programming and setting closely integrated	programming in the hands of a programming/planning department, and differentiation between operating setting and programmer skills

Complexity of component

A technologically advanced horizontal machining centre, with automatic tool-changing facilities, was usually manned by a skilled operator because of the high costs involved in the machine itself, the necessary tooling, and frequently the nature and complexity of the components manufactured. Apart from any 'micros' or planning sub-routines, these machine types have increasingly sophisticated input systems which substantially simplify programming procedures. Thus, in this situation, the differentiation between planning and operating may not be particularly polarised. By contrast, on a simpler milling machine with less complex components, the program is probably also less technically demanding, but the operator may only be semi-skilled. In this instance, skill polarisation may be enhanced.

The time needed to make a program is the central factor associated with different allocations of programming related functions; the more time it takes to make a program, the less is programming likely to occur on the shop-floor. However, such a relationship presupposes that an operator cannot make a program without the machine being idle. But this is only true to a limited extent. We have seen, for example, cases of operators drafting programs after working hours at home. It is also possible, on some machines, to compile the program for the next job while the present one is being processed.

National institutions

National differences interact with company and batch size differences: in Germany, the similarity between organisation, labour and technical practices of small and larger companies was greater than in Britain, where there appeared to be a split between pragmatically flexible small plants and organisationally more segmented larger plants. Formal engineering qualifications at various levels were relatively more common in Britain in larger plants, whereas often they were not represented in small British plants. In Germany, by contrast, formal qualifications were common to small and larger plants alike.

In both countries, CNC operation was generally seen as exacting less 'informatics' skills than advanced machining talents(12). Programming aids on the machine or in the planning departments are seen to be tools of increasing facility to control a process which has become ever more demanding from the point of view of precision, machining speeds, tools, fixtures and materials.

Market conditions

During the study it became evident that changes in competitive and marketing strategies were taking place and were having a marked effect upon production and other policies. Enterprises both in Britain and in Germany, stated that they were, indeed, being forced, increasingly to cater to small market 'niches' rather than for homogeneous mass markets. More individualised, customised products, with a greater number of product variations, were seen as being required. Product innovation, as part of this development, was also leading towards increased component complexity, and as a consequence, increased financial costs associated with work-in-progress and finished stock.

Thus both market-oriented as well as financial considerations point in the direction of smaller batch sizes and more frequent conversion; this has important socio-technical implications. It is through these developments that we can see the effect on skills. The increased variability of batches is not one which can be handled bureaucratically through a conventional increase in the division of labour, and in turn mitigates, or even possibly reverses, the tendency to skill polarisation. It implies increased flexibility right at the level of the machine and the operator. Every CNC operator is likely to have to deal with a greater and more frequently changing range of jobs; part of this is related to the increased sophistication of the machine control system through which more flexible change-overs and improvements of programs can be achieved. The crucial 'bottle-necks', however, may not be information-processing and calculating skills; experience indicates that the most crucial problems refer to tooling, materials, feeds, speeds, faults, and breakdowns. Skills in handling these problems are most directly developed on the machine. Thus, while programming skills are required, increasing emphasis must be placed on the maintenance of 'craft' skills on the shop-floor, rather than the converse (as implied by the 'pessimists' writing on the subject).

Companies, particularly in Germany, are increasingly seeing the merits of stressing craft skills. This de-polarisation of skills and qualifications structures is a viable option which falls within the present logic of CNC development and application. This is not because it is a necessary consequence, but because CNC has beean developed in a context which links economic success with de-polarisation. There is a striking kinship between CNC and some of the craft trades, and the renewed interest in companies, again particularly in Germany, to train and employ skilled workers.

Increased programming or program-changing in the workshop may thus blur status boundaries for blue as well as white-collar workers. It would, however, be misleading to interpret this as another step towards the 'post-industrial' society, as 'information-processing' work or as a 'service' function, as so often happens. Whilst it is true that workers are dealing with increasingly sophisticated information technology, this may only concern the *tools of their trade* rather than their *working goals*.

CONCLUDING REMARKS

To sum up, previous manufacturing applications have been to date essentially geared to specialised, homogeneous mass markets, inflexible automation, an erosion of craft skills, and an increased emphasis on separate planning and programming activities. In the labour market, there was also a move towards white-collar occupations, information-processing, administration, and different kinds of clerical work, particularly visible in the British experience. The socio-economic context of CNC application, however, may reverse this trend; there is now increasingly a focus on craft skills and the levelling-out of the growth of indirectly-productive employees, very clearly seen in the German firms studied. This, however basically follows the organising tradition of the respective national work-cultures. It is misleading to examine technological change outside this 'societal' context; rather, we see it interacting with it. In so far as the study has implications for organisational theory, it must be seen as offering support for the view that technological change needs to be seen as having much more open social implications and leaving organisation designers more options than hitherto assumed.

ACKNOWLEDGEMENT

We gratefully acknowledge the support of the Anglo-German Foundation for the Study of Industrial Society for this study, as well as the help of CEDEFOP, the European Centre for Vocational Training. For professional dialogue and criticism, we have the pleasure to thank Marc Maurice, Francois Eyraud, Donald Gerwin, Georges Dupont, Michael Fores, Derek Allen and Jonathan Hooker.

REFERENCES

1. BRAVERMAN, H. Labor and Monopoly Capital; The Degradation of Work in the Twentieth Century. New York: Monthly Review Press, 1974.

2. TRIST, E. The Evolution of Socio-technical Systems; A Conceptual Framework and an Action Research Program. Occasional Paper, No. 2. Ontario: Ontario QWL Centre, 1981.

3. PRAIS, S.J. Vocational Qualifications of the Labour Force in Britain and Germany. London, National Institute of Economic and Social Research, discussion paper 43, 1981.

4. SWORDS-ISHERWOOD, N. and SENKER, P. Social Implications of Automated Small Batch Production. in: UTOMATED SMALL BATCH PROCUCTION COMMITTEE: Automated Small-Batch Production: Technical Study. London: Department of Industry, 1977.

5. JONES, B. Destruction or Re-distribution of Engineering Skills? The Case of Numerical Control, in WOOD, S.(ed) The Degradation of Work? London: Hutchinson, 1982.

6. MARSH, A., HACKMANN, M., and MILLER, D. Workplace Relations in the Engineering Industry in the UK and the Federal Republic of Germany. London: Anglo-German Foundation for the Study of Industrial Society, 1981.

7. SORGE, A. and WARNER, M. Manpower Training, Manufacturing Organization and Workplace Relations in Great Britain and West Germany. British Journal of Industrial Relations, 18, 1980, p.318-333.

8. MAURICE, M., SELLIER, F. and SILVESTRE, J.J. La Production de la Hierarchie dans l'Enterprise: Recherche d'un Effet Societal. Revue Francaise de Sociologie, 20, 1979, p.331-365.

9. MAURICE, M., SORGE, A. and WARNER, M. Societal Differences in Organizing Manufacturing Units: A Comparison of France, West Germany and Great Britain. Organization Studies, 1, 1980, p.59-86.

10. SORGE, A., HARTMANN, G., WARNER, M. and NICHOLAS, I. Micorelectronics and Manpower in Manufacturing: Applications of Computer Numerical Control in Great Britain and West Germany. International Institute of Management, West Berlin: IIM, Research Report IMP 81 16, 1981. (a revised version is to be published by Gower Press in mid 1983.) It is also published in German under the title Micro-electronik und Arbeit in der Industrie, Stuttgart Campus Press, 1982.

11. GOHREN, H. Konstruktive und Steuerungstechnische Entwicklungen bei Maschinen fur die Mittelserienfertigung. VDI-Z 121: 1979, p. 403-409.

12.LEE, D.J. Skill, Craft, and Class; a Theoretical Critique and a Critical Case. Sociology, 15: 1980, p. 56-78.

Technical training and technical knowledge in an Irish electronics factory

Peter Murray and James Wickham

A central theme in the discussion of information technology is its impact on skill levels within the workforce. This paper examines one aspect of this question: the use of technical training and technical knowledge in the manufacture of the actual hardware of information technology. Information technology is often presented as 'requiring' greater training or 'causing' a more skilled workforce. However, using a case study of a US-owned electronics in the Republic of Ireland(1) This paper argues against such technological determinism.

The first part of the paper describes the production process in the case study factory and relates this to the international distribution of skills within the electronics industry. The second part examines how training and education link access to jobs in the factory to the wider social structure. The final part shows how technical knowledge within production is shaped by both the (sometimes contradictory) requirements of management and the conflicts of interest between management and employees.

AN ELECTRONICS FACTORY IN THE INTERNATIONAL DIVISION OF LABOUR

The case study factory is located in a new industrial estate in North Dublin and employs about 125 people. It is a wholly owned subsidiary of 'Hightech International' (the name is a pseudonym for all the usual reasons). Hightech International is a relatively small American electronics corporation, with two manufacturing plants in the USA and a European sales headquarters near London. The corporation itself is only about 15 years old and the Dublin factory began operations within the last five years.

Hightech's Dublin factory assembles for the European OEM (Original Equipment Manufacturer) market the firm's main product, a mini-computer bought by other manufacturers as a building block in systems designed for particular applications. It also manufactures small quantities of two other products: an office automation system and an ATE (automatic testing equipment) system used in the testing of assembled printed circuit boards. Hightech set up in Dublin not because of any need for cheap labour per se (as is the case with US 'offshore assembly plants in South-East Asia) but because it required a final stage manufacturing facility within the protected European market. The printed circuit boards used in the plant are manufactured by Hightech in the USA. Other components are imported from the USA through the parent company, or are sourced from the Dublin plant following approved specifications laid down at the corporation's headquarters in 'Chipville' California.

Table 1: Hightech workforce: job type by sex

	Male	Female	Total
Assembly (1)	19	20	39
Test Operator	10	-	10
Technician	18	-	17
Clerical	-	17	17
Lower Management (2)	22	-	22
	68	37	105

(1) Assembly includes electronic and electro-mechanical assemblers, reworkers, quality control inspectors, storekeepers and ancillary general workers.

(2) Lower Management includes production supervisors, engineering and quality control technical support staff, materials and production planners and accountants.

Source: Workforce Survey

About two thirds of all employees in the Dublin plant work in direct production on the large open-plan factory floor (Table 1). The largely female electronics assembly workers work on two lines: a small line producing cables and the main line where the printed circuit boards (PCBs) are assembled. Here the individual components are prepared ('prepping') and assembled onto the PCBs ('stuffing'). After these two manual stages the boards are passed through an automatic flow solder machine and then go to the 'touch up' area for visual inspection and the removal of solder splashes etc. They are then tested for faults, firstly by automatic testing equipment operated by (male) test operatives and then by diagnostic testing in which (male) technicians check for systems faults. Any faulty boards are passed to reworkers (another group of assemblers) for repair. The boards are then placed in the 'burn in' environmental test chamber where programs are run on them at very high temperatures to simulate the operating conditions under which component failure is most likely to occur. Finally, the boards are configured by test operatives and technicians, mounted in chassis and equipped with power supply, consoles, etc. by the (male) electro-mechanical assembly workers. The completed product then goes to the goods outward store prior to shipment to customers within the EEC.

Table 2: Irish trade in electronics (selected categories), 1980

ISIC Classification		EEC9 (£m)	Great Britain (£m)	USA (£m)	Total (£m)
75123 cash registers incorporating a calculating service	Exports	5.4	3.0	-	6.5
	Imports	1.7	0.9	-	2.3
	Balance	+3.7	+2.1	-	+4.2
75220 complete digital ADP machines (CPU) with input and output devices	Exports	84.8	45.3	2.2	100.9
	Imports	5.7	4.2	23.0	29.0
	Balance	+79.1	+41.1	-20.8	+71.9
75250 peripheral units including control and adapting units	Exports	38.7	10.2	0.3	47.2
	Imports	6.1	4.6	4.6	18.5
	Balance	+32.6	+5.6	-4.3	+28.7
75990 parts, NES, of machines 75121-75128 and 75210-75280	Exports	62.4	21.9	4.6	76.9
	Imports	27.6	15.5	67.8	100.5
	Balance	+34.8	+6.4	-63.2	-23.6
75 office machinery and automatic data processing equipment	Exports	209.7	85.3	7.3	257.9
	Imports	60.6	36.2	101.4	135.9
	Balance	+149.1	+49.1	-94.1	+122.0

Source: Central Statistics Office, Trade Statistics of Ireland.

This description makes clear that the factory is an integral part of a production process that is organised on an international scale even within such a relatively small corporation. In this it is typical of the US-owned electronics firms that dominate the 'Irish' electronics industry. A postal survey carried out early in 1981 in which 93 of the 100 manufacturing firms in the industry at that date participated showed that 42% of electronics plants in the Republic were US-owned and that they employed 63% of a reported workforce of 11,300(2). As Table 2 shows, the industry imports components from the USA and exports finished products to Europe. This particular place within the international division of labour means that any explanation of skill levels within Irish electronics plants in terms of the importance of research and development (R&D) in the industry as a whole is completely inaccurate. Like nearly all US electronics firms in Ireland, Hightech concentrates its basic research in the USA(3). Not surprisingly therefore, the distribution of skills within the US and Irish electronics workforces is very different. As Table 3 shows, the USA has a noticeably higher proportion of technicians and professionally qualified employees than does Ireland.

At the same time this hardly means that all workers in Irish 'branch plants' are semi-skilled workers. Table 3 also shows that the Irish electronics industry has a higher proportion of technicians and professionals than the rest of the Irish manufacturing industry. The example of Hightech shows why this occurs. Firstly, like most US electronics factories in Ireland, the Dublin plant is the final manufacturing stage for products destined for the European market. Consequently, all final stage testing has to occur in Dublin. It is the importance of testing, rather than of R&D, within Irish electronics manufacture that explains the high proportion of technicians. It has also suited Hightech to locate in Dublin the detailed adjustment of its product to European technical standards, which again requires more skilled workers, even though such minor 'product modification' can hardly count as R&D in the normal sense of the word(4). Secondly, like many but not all US firms in Ireland, the factory's manufacturing process is batch rather than mass assembly. The basic product has an extremely wide range of options, so that approximately 200 different types of board are assembled, usually in batches of between 25 and 50. This again increases the proportion of administrative, clerical, supervisory and testing personnel within the workforce(5). It is these two features - final stage manufacturing and the predominance of batch production - which differentiate the workforce of Irish branch plants within electronics from the offshore assembly factories of South-East Asia, where up to 80% of the total workforce is involved in assembly work. The skill structure of the Irish electronics industry, in other words, is in the first instance the result of its intermediate position within an international production system(6).

EDUCATION, TRAINING AND ACCESS TO JOBS

The discussion of the educational requirements of the new technology has largely ignored a central theme in the sociology of education: that of class inequality in access to education. By contrast

Table 3: Occupational structure of US electronics industry, Irish electronics industry, all Irish industry (various years).

	US electronics (1980) %	Irish electronics (1981) %	All Irish industry (1976) %
Managers	11	6.4	6.5
Professionals	17	5.3	1.3
Administrators	na	3.9	5.2
Technicians	11	7.8	1.6
Supervisors	na	5.1	4.7
Sales	1	na	na
Clerical	12	7.5	7.7
Craft workers	10	3.1	12.6
Apprentices	na	-	4.2
Operatives etc(1)	32	57.4	56.3
Labourers etc(2)	4	3.2	na
Total	100	99.7	100.1
N		11,338	220,500

(1) 'Operatives' (USA); 'Non-craft production workers' (Irish electronics); 'Other workers' (all Irish industry).

(2) 'Labourers' (USA); 'Others' (Irish electronics).

Source: US electronics: Global Electronics Information Newsletter, September 1982; Irish electronics: Postal Survey, 1981; All Irish industry: AnCO, Research and Planning Division, Manpower Survey 1976.

in this section we examine how access to jobs at Hightech is connected to the wider social structure through eductional and training requirements.

Table 4: Hightech workforce: Job type by educational achievement.

Highest examination passed	Assemblers	Test operatives	Tech-nicians	Cler-ical	Lower managers and professionals
None	10	-	-	2	2
Group	5	-	-	-	-
Intermediate	5	-	1**	6	3
Leaving	10	10	1**	9	5
Higher	-	-	15	-	11
Total	30*	10	17	17	22

'Group', 'Intermediate' and 'Leaving' are the main Irish educational certificates at secondary level. A small number of workers were educated in Britain. In these cases, 'O' and 'A' levels have been treated as equivalent to the Leaving Certificate; CSEs are treated as equivalent to the Group Certificate. 'Higher' includes all third-level examinations, including certificates and diplomas as well as degrees.

* Excludes all assemblers over 31, most of whom were married women with children, who had returned to industrial employment after a long break, and a small group of political refugees working in the factory.

** The two technicians without third level qualifications had been promoted from test operative and were studying part-time for a third level qualification.

Source: Workforce Survey.

Entry to the technicians' jobs requires completion of either two or three years full-time third level technical education(7). Of all the main groups of employees at Hightech, including the lower managers, the technicians had the highest educational qualifications (Table 4). Given that 40% of them were also recruited direct from college and that for all of them Hightech was their first permanent job, they were also the youngest group within Hightech's generally young workforce (Table 5).

Table 5: Hightech workforce: job type by age group

	16-24	25-30	31-40	41-50	51-60	Total
Assembly	24	8	3	2	2	39
Test Operative	9	1	-	-	-	10
Technician	15	2	-	-	-	17
Clerical	14	1	1	1	-	17
Lower Management	6	9	3	3	1	22
Total	68	21	7	6	3	105

Source: Workforce Survey

The requirement of a third level educational qualification by the technicians has two consequences. Firstly, they are more likely than clerical workers, and in particular than assembly workers, to have been born outside of Dublin, which is consistent with the finding that much of the Dublin 'petty bourgeoisie', unlike the manual working class, is recruited from outside the city(8). Secondly, given the well-documented inequality of educational opportunity in Ireland(9), access to their jobs is linked to class of origin. They are more likely than other groups, again including the lower managers, to have come from at least lower white-collar backgrounds (Table 6).

In other countries technicians have often been recruited via apprenticeship and subsequent promotion, so that technicians are upwardly socially mobile: in their national sample survey of technicians in British industry, Roberts et al found that approximately 50% came from manual working class backgrounds, and in his study of two electronics factories in Milan, Low-Beer reports 49% of the technicians being of working class origins(10). Jobs in Irish electronics by contrast appear to provide no channel of occupational mobility to counter-balance the restricted entry of working class children to even non-degree level third level education. Finally, since virtually no women take technicians' courses at third level, the qualification requirement by itself ensures that the technicians are all men(11).

Table 6: Hightech workforce: social origins

Father's main occupation	Assemblers	Test operatives	Tech- nicians	Cler- ical	Lower management and professionals
Higher professional	-	-	3	-	2
Lower professional	-	-	-	1	1
Proprietor	4	4	5	4	5
Farmer	2	1	2	-	-
Intermediate Non-manual	5	-	3	1	5
Skilled manual	11	3	2	6	2
Service	4	-	-	1	3
Semi-skilled manual	9	1	1	3	2
Unskilled manual	1	1	-	1	-
Total	36	10	16	17	20
Missing observations	3		1		2

Source: Workforce Survey

Over 80% of the technicians considered that their formal education not only had been important for gaining jobs at Hightech, but also provided 'basic knowledge' needed in the job. The situation is very different for the assemblers, whose job requires no particular educational qualification. However, over a third of assemblers have attended a brief six weeks course in electronics assembly run by AnCO (the Irish Industrial Training Authority) and in all three quarters have had some form of training since leaving full-time education. While those assemblers with some training were almost as likely as the technicians to consider that this qualification had been important in *acquiring* their jobs, a much smaller proportion considered that it helped them to do their job. This applied just as much to those who had done the electronics assembly course as to those who had apparently less relevant qualifications (secretarial, nursing, etc.).

How such apparently irrelevant training can be important becomes clearer when the recruitment of assemblers is examined more closely. Today in Ireland there is a growing stress on the need to prepare school leavers for industrial work and indeed in 1981, 40% of males and 21% of females who had left school a year earlier and who were in full-time employment were employed in industry(12). In Hightech however nearly all the young assemblers had started work in the service sector and most had no previous industrial experience. The small number who had already worked in factories tended to be recruited to Hightech direct from the labour market(13). By contrast, those with no industrial experience were recruited in one of two ways. Firstly, some were placed in their jobs by the training authority after they had completed the electronics assembly course. Secondly, the remainder gained their jobs through personal contacts, having either friends or relatives already working in the factory. That these two methods of recruitment were alternatives suggests that the importance of training, even in electronics assembly, is not that it provides the minimal skills needed in work, but rather that it is a mode of access to work(14).

Educational standards also influence access to the assemblers' jobs. Whereas of all 1980 school leavers who were employed in skilled or semi-skilled manual work in 1981, only 4% of the girls and 30% of the boys had passed the Leaving Certificate(15), at Hightech the proportions amongst the young assemblers are 21% and 42% respectively. And these relatively high educational standards are deliberate policy on the part of the firms: the Engineering Manager in particular stressed to us that he preferred to have assemblers with the Leaving Certificate on the shop floor.

Despite this, the actual assembly jobs are designed to require as little skill as possible. Assembly work involves following colour codes on documents produced by the Engineering Department:

> 'We produce a range of seven documents which in theory we aim to have idiot-proof: documents such that a girl can assemble in a very complicated printed circuit board, can do the touch up work on it, add the odd components, put them in the right way... To allow the board to be manufactured by people as unskilled as the people who are in the assembly area we've got to produce very sophisticated documentation which means we take the engineering down step by step to a simplified level that the operator can understand'. (Engineering Manager)

Yet assembly work in Hightech, as indeed in most Irish electronics factories, hardly conforms to the usual image of women's semi-skilled work presented in recent literature(16). Firstly, although boards pass from one work station to another along a line, there is no conveyor belt and hence no machine pacing. Secondly, supervision and quality control relies largely on the testing carried out by the test operatives and technicians, rather than on any close control by immediate supervisors.

Thirdly, there is no piece-rate system. Finally, production is batch production with in addition continual minor changes in the design of each board. Consequently, while requiring no technical skill, the work requires what Offe defines as 'normative orientations': internalised commitment both to the particular work and to the authority structure of the firm(17). Workers have to be motivated to anticipate what is the 'correct' behaviour in a situation that cannot be completely controlled by supervisors. For the assemblers therefore, technical training, being spoken for by friends and relatives, and educational qualifications, all ensure that those who are recruited can be assumed to be 'reliable' or 'responsible' in the way that management defines these terms.

By contrast with both technicians and assemblers, internal promotion linked to directly useful technical training is important for the small group of test operatives. The first test operatives at Hightech had a two year technicians' certificate and were all subsequently promoted to technician. The current group acquired their jobs either by a six-month AnCO electronics training course before joining Hightech or by promotion from the shop floor after part-time study. Since all hope to continue part-time education and to become technicians themselves, they would appear to be on a social mobility route which leads from assembler through to technician.

Certainly the current test operatives are closer to the assemblers than the technicians in terms of their social origins (Table 6). However, test operatives are distinguished from the assemblers by the fact that all of them have passed the Leaving Certificate. Promotion to test operative is therefore linked to achievement at school, and is not dependent solely on steps taken after entry to full-time work. Furthermore, this mobility channel is not open to all young assemblers: whereas about half of the assemblers are women, all the test operatives are men. Women assemblers were in fact interested in promotion to test operatives jobs, but were apparently unwilling to attend part-time courses in the city centre in order to do this(18).

Partly in order to meet this problem, while we were carrying out fieldwork the firm introduced an evening course in basic electronics theory for interested assemblers. This would have opened up the test operative grade to women for the first time. The course evoked considerable enthusiasm, but was quickly discontinued when the current recesion forced retrenchment in the firm. Finally, as the next section will show, any further expansion of the test operative grade is dependent on changes within the technician grade - changes which would create here a clear hierarchy with very restricted internal promotion. In other words, promotion from assembler to test operative is decreasingly a step on a wider promotion ladder.

During the reform of Irish education in the 1960s a 'human capital' approach explicitly linked the 'modernisation' of the curriculum with increased equality of opportunity, even if the latter aim was not actually achieved(19). As we have however shown elsewhere, today the increasing concern with technological education has obliterated from public policy any such concern with equality(20). The class origins of Hightech's technicians suggests that in Ireland the growing importance of full-time education for providing technical qualifications also reinforces the inheritance of class inequality.

Technological education obviously involves implicit ideologies, but the apparently self-evident usefulness of such education both to those who receive it and to the society at large means that, in the case of the technicians, its role in the transmission of inequality remains unquestioned. By contrast, the ideological elements in the much briefer technical training received by assemblers are more apparent, since its technical uselessness in their work is quite clear to them (if not to outsiders). This is in line with Offe's general argument that today in industrial work training, recruitment and promotion are presented as involving the mastery of technical rules, but actually primarily concern the internalisation of normative orientations or ideology(21). Indeed, a common finding of recent industrial sociology is that employers hiring manual labour use 'skilled' as a synonym for 'responsible'(22). However, such studies have been concerned only with male workers, and so by default women remain seen as 'unskilled', even if skill is reformulated to mean exposure to ideological selection. Consequently, discussion of the 'suitability' of women for semi-skilled assembly work has focussed on sexual stereotypes and their relationship to the domestic division of labour(23). However, the selection of women assemblers at Hightech shows that at least in electronics manufacture in a situation such as Ireland, semi-skilled assembly work involves other, and more work-related, 'normative orientations'.

TECHNICAL KNOWLEDGE AND THE ORGANISATION OF PRODUCTION

Unlike the assemblers, the technicians do use their technical training in work. Yet as this section will show, the precise way in which this occurs is the result of a series of social processes and not simply of Hightech's production technology.

The wide product range and frequent minor changes of production at Hightech mean that the technicians need to be enthusiastic and relatively well educated:

> 'If things don't go right and you just haven't got all the training requirements and equipment in place, (the technicians) just merrily work away and do the best they can... so that's a big advantage and that's what helps us to have such a wide product range'. (Engineering Manager)

This technical enthusiasm is maintained by a challenge such as the introduction of a new manufacturing process:

> 'The technicians and the engineers just tear into it... reading it at home and like they're just very interested in it technically. A lot of these guys still have the college zest for learning, you know, and there's a bug in the electronics business in terms of the academic side of it. People just want to know more, there's just

> no limit to the knowledge you can get, especially from the technicians on the shop floor'. (Engineering Manager)

However, such enthusiasm and high educational standards can be unproductive for the routine work of the technicians, namely fault finding. Technicians are supposed to normally observe a maximum of two hours per board for fault finding, and if the fault cannot be located within this time the board is meant to be put aside for a possible slack period. Yet Hightech's Dublin technicians, unlike their equivalents at Hightech's Chipville plant, are unwilling to restrict themselves like this:

> 'In Chipville the technicians are trained from a very low academic background, very low interest in it, and they can do their job more efficiently than our guys because they know no better. If the board doesn't work then they just have to give it to somebody else. Our guys, Jesus, some of them out there will practically dissect the chips to see what's wrong with them which is totally ineffective and non-productive. They're just not suited to assembly line work which is really what it is in a bit more of an elaborate stage'. (Engineering Manager)

Thus the firm requires its technicians to be able to work independently <u>and</u> to tolerate routine work. This contradiction was to some extent related to a conflict between the Personnel and the Production Departments over recruitment. Whereas Personnel considered that technicians should be hired at the lower end of the going salary rates, Engineering wanted better technicians and felt that the firm should be prepared to pay for them:

> 'Your Personnel Manager or your Financial Controller or your Managing Director, you can't expect them to be into it in the same way. Take Personnel - all they can do is say "Well, look Apple make computers, Amdahl make computers, we make computers". (But) I have to have the best of the technical people just coming out of the colleges to work this product and get it working properly because our product is more sophisticated. You do things with our computers you don't do with most people's computers'. (Production Manager)

How the technicians' knowledge is used is also shaped by a conflict of interest between them and management. For the technicians, Hightech is simply the starting point of a career in electronics. In a situation of apparent shortage of technicians in Irish electronics plants, they think of themselves as 'job hoppers'rather similar to original computer programmers of the 1960s described by Kraft(24). Such movement appears feasible since they are all young and single. Of the five groups of employees, they were least likely to expect to be in Hightech in three years' time, but also most likely to expect to be both in electronics and in a higher level job.

These mobility aspirations mean that they are prepared to tolerate what they saw as the low technicians' salary at Hightech. The firm's wide product range and particular its manufacture of the commercial system and automatic testing equipment promise experience which would be useful to gain better jobs in other firms. Their interests lie in developing a general technical knowledge through which allowed them to amass a wide product knowledge.

By contrast, management is interested in utilising general technical knowledge only in a supervisory capacity. Hence it wishes to replace the informal consultation that currently goes on amongst technicians in the testing area with a clear hierarchy based on a new higher 'C' grade technician:

> 'This tech C is a sort of super tech if you like. What I want is a sort of electronic guru. He sits in the middle of an area - OK he's not always sitting there, he's actually walking around looking at these other technicians who are working - and when they've got a problem he just lays a magic hand on their shoulder and says 'I C 56 for these reasons...'and the Technician A is enlightened and really has learnt something'. (Production Manager)

While for most technicians, their work would be narrowed and routinised:

> 'My aim is to have a very, very small core and then have a whole gang of people. I don't mind, they can be here for six months or two years, they can do their thing... If you look at the structure of the technical people - test ops to sophisticated technicians - as being pyramid shaped, I'm trying to expand the base of my pyramid and bring down the width of the top'. (Production Manager)

In such a situation most technicians would develop only a narrow product knowledge, and their general knowledge would be very restricted. The increased number of test operatives using the ATE would further reduce 'unnecessary' and in fact counter-productive technical expertise:

> 'If (the board) didn't go through automatic testing and went to a systems testing (the technician) puts a diagnostic (program) through the machine and it fails. So the readout he has is not 'chip number 17 in backwards'. It says that the function of the machine... fails to carry all the data across. He looks at the diagnostic and he says, 'Well, why should that happen now? Maybe the signal should be a different signal coming out of that part of the board'. And he's much more into an application fault and all the problem is is that the chip is in the wrong way. But he's going to have to go through all this systems type thinking to figure that out'. (Engineering Manager)

Increasingly boards are designed to utilize ATE to the full. Although dramatic developments, including the introduction of robots, have been predicted in electronics assembly(25), it is in fact in testing that automation is advancing most rapidly. The design of the boards is now interwoven with the development of the software to test them on ATE (and in Hightech both are carried out in Chipville rather than Dublin). Management in Hightech hope that this will enable the test operatives' work to become closer to machine minding work and less like technicians' work.

As far as the technicians are concerned, their jobs have already become narrower. When the Dublin factory opened, the technicians were involved in almost all aspects of production:

> 'When they started this place up, they hired ten technicians and then another ten technicians and they all went off to California for three months and had all this product training. And when they came back, there was the Materials Manager and there was nobody else... There was product coming in through the back door and these technicians were taking it in, cutting off the goods received note, unboxing it, taking it out, testing it, ... putting it in the burn-in room, taking it out, fixing it, putting it down into finished goods, pulling it out again, putting it in a configuration test... This was all technicians'. (Production manager)

But as the immediate start-up phase ended:

> 'These technicians started to be boxed-in. As the company expanded the jobs that the technicians were doing became less and less, and they were coming to do what we would understand to be a technician's job - being shackled to his work bench almost, testing a single board or a single product line'. (Production Manager)

The result of these changes is that the technicians' mobility aspirations are already proving to be unrealistic. A technician requires about a year to develop detailed product knowledge necessary to effectively test the particular boards being produced at Hightech, and it is at this stage that he is most valuable to the company. However, once he has achieved this he starts to become bored with the work. The company may be able to hold on to him by promoting him to technician 'B', but soon he will be looking for another job in a different company. Many of the technicians hope for promotion to managerial or professional engineering grades and are continuing their studies part-time to achieve this. However, here they will be competing with the graduate electronics engineers that the Irish educational system is producing in ever-greater numbers(26). Consequently, technicians already move sideways between firms, rather than upwards as they hope:

> 'It's quite normal to find a technician who let's say has been graduated seven years and he has spent two years with three companies and he's now on his fourth company. At the end of the day that technician will be a technician just as the Post Office has had technicians for fifteen or twenty years at the same level. But in the public debate we have elevated those to a much higher level than is real'. (Personnel Manager)

This blocked mobility has been noted as characteristic of British engineering technicians(27). The distinguishing feature of Irish electronics technicians however is that, as yet, they remain unaware of the limitations of their situation.

The use of technical knowledge in the factory involves contradictions and clear conflicts of interest. Management requires from most technicians a developed but narrow product knowledge, and from a few a broader general knowledge used mainly in a supervisory capacity. By contrast, technicians hope to develop a wide product knowledge and, in addition, to have the chance to use general technical knowledge in their work. It is within this context that management plans to extend the hierarchy amongst the test operatives and technicians and so redistribute technical knowledge within the factory have to be interpreted.

CONCLUSION

In the Irish situation, even branch plant assembly requires some technical knowledge in the workforce. Yet how this knowledge, produced by state training and educational institutions, is actually used in production is not unproblematic. Both assemblers and technicians receive technical training, though obviously of very different lengths. In the case of the assemblers their training is largely a way in which the state training system screen potential employees on behalf of the firm. In the testing area, where technical knowledge aquired in education does play an important role in the work situation, how it is used and developed is the result of the interplay between the conflicting needs of the technicians and the firm. Even in high technology industry - and perhaps particularly there - it is hardly only the technology that determines the nature of technical knowledge.

REFERENCES

1. Fieldwork was carried out in 1981/2 as part of a wider project on the electronics industry in the Republic of Ireland on which we are currently engaged. The project is funded by the National Board for Science and Technology and the Employment Equality Agency. The case study involved a survey of all employees in the factory with the exception of senior management, detailed observation of the main jobs and extended tape-recorded interviews with management and trade union representatives. We would like to thank all who participated in the study for their co-operation.

2. Further results of this survey are discussed in MURRAY, P. and WICKHAM, J. Technocratic Ideology and the Reproduction of Inequality: The Case of the Electronics Industry in the Republic of Ireland. in DAY, G. *et al*. (eds.), *Diversity and Decomposition in the Labour Market*. Aldershot: Gower, 1982, p.179-210.

3. SCIENCE POLICY RESEARCH CENTRE. *The Irish Electronics Industry: A Review of Structure and Technology with Policy Implications*, August 1981, (mimeo), University College Dublin, pp.138ff.

4. In the case of Hightech this 'R&D' appeared to largely involve modification to conform with European electricity supply.

5. It has been argued that even South-East Asian 'offshore assembly' electronics plants require quite complex management, given the frequency of product changes. LESTER, M. The Transfer of Managerial and Technological Skills by Electronic Assembly Companies in Export-Processing Zones in Malaysia. in SAHAL, D. ed., The Transfer and Utilization of Technical Knowledge. Lexington, Mass: DC Heath, 1982, p.209-224.

6. This argument is developed with extensive comparative data in O'BRIEN, R. Electronics Industry Development in Small European Countries and the NICs in South-East Asia. Paper read to the Development Studies Association Conference, Long Waves, Technology and International Development, Edinburgh, April 1983. Accounts of multi-national branch plants in Ireland in terms of the 'new international division of labour' claim that the main reason for location in Ireland (and other peripheral regions of Europe) is cheap labour; in fact the main reason is access to the EEC market, and labour cost considerations follow from this in choosing between European regions. See FROEBEL, V., HEINRICHS, J. and KREY, O. The New International Division of Labour: Structural Unemployment in Industrialised Countries and Industrialisation in Developing Countries. London: Cambridge UP, 1980, and the review by JACOBSON, D., WICKHAM, A. and WICKHAM, J. Capital & Class, no. 7 Spring 1979, p.125-30.

7. Either a two-year technicians' certificate from a Regional Technical College or a three-year City and Guilds diploma from a Dublin College of Technology.

8. HUTCHINSON, B. Social Status in Dublin: Marriage, Mobility and First Employment. Dublin: Economic and Social Research Institute, 1969.

9. ROTTMAN, D and HANNAN, D. The Distribution of Income in the Republic of Ireland: A Study in Social Class and Family Cycle Inequalities. Dublin: Economic and Social Research Institute, 1982. This study concludes (p.63) that access to university education in Ireland is more inegalitarian than in the UK.

10. ROBERTS, B. et al. Reluctant Militants: A study of Industrial Technicians. London: Heinemann, 1972, p. 130; LOW-BEER, M. Protest and Participation: The New Working Class in Italy. London: Cambridge UP, 1978, p. 166.

11. In 1980 women comprised 5% of the 514 students studying electrical and electronic engineering in Regional Technical Colleges and 3% of the 539 studying in Colleges of Technology. DEPARTMENT OF EDUCATION, Statistical Report 1979-80. Section VII, Table 3.

12. NATIONAL MANPOWER SERVICE. School-Leavers 1980: Results of a Survey carried out in May 1981. Dublin: Department of Labour, 1982, Tables 7 and 8.

13. They mostly either applied directly to the company themselves or were recruited through the National Manpower Service (the state employment agency).

14. For a general account of the expansion of technical training in Ireland in these terms, see WICKHAM, A. Industrial Transition and Training Policy in the republic of Ireland. In: M. KELLY, L. O'DOWD and J. WICKHAM, (eds.). Power, Conflict and Inequality. Dublin: Turoe Press, 1982, 147-58.

15. NATIONAL MANPOWER SERVICE. School Leavers. Tables 9 and 10. The Irish Leaving Certificate is the final school examination.

16. See, for example, CAVENDISH, R. Women on the Line. London: Routledge and Kegan Paul, 1982. This issue is discussed in more detail in MURRAY, P. and WICKHAM, J. Industrial Relations in Irish Electronics Branch Plants. Paper presented to the British Sociological Association Annual Conference, April 1983.

17. OFFE, C. Industry and Inequality: The Achievement Principle in Work and Social Status. London: Edward Arnold, 1976.

18. Since none of the young women assemblers had their own children and attendant childcare commitments, one suspects that they were also put off by the male culture of the part-time courses.

19. On the objectives of these reforms, see CRAFT, M. 1976, Economy, Ideology and Educational Development in the Republic of Ireland. Administration 18 (4), Winter 1970, 363-74. WICKHAM, A. National Education Systems and the International Context: The Case of Ireland. Comparative Education Review, 24, 1980, 323-337. On the small impact these changes have on equality of opportunity, see especially TUSSING, D. Irish Educational Expenditures - Past Present and Future. Dublin: Economic and Social Research Institute, 1978.

20. MURRAY and WICKHAM, 1982 op.cit.

21. OFFE, op cit.

22. BLACKBURN, R. and MANN, M. The Working Class in the Labour Market. London: Macmillan, 1979, esp. pp 102-109; OLIVER, J. and TURTON, J. Is there a Shortage of Skilled Labour?, British Journal of Industrial Relations, 20 (2), July 1982, 195-200.

23. PHILLIPS, A. and TAYLOR, B. Sex and Skill:Notes Towards a Feminist Economics. Feminist Review, no.6, 1980, 79-88. ELSON, D. and PEARSON, R. Nimble Fingers make Cheap Workers: An Analysis of Women's Employment in Third World Export Manufacturing. Feminist Review, no.7, 1981, 87-107.

24. KRAFT, P. The Industrialization of Computer Programming: From Programming to Software Production. In: ZIMBALIST, A. (ed). Case Studies on the Labor Process. New York and London, Monthly Review Press, 1979, p. 1-17.

25. Cf. Global Electronics Information Newsletter. (Pacific Studies Center, Mountain View, California), no. 11, June 1981, and no. 25, October 1982.

26. The main policy priority for higher educational investment in the 1980s is identified by the White Paper on Educational Development (Prl. 9373, Stationery Office, 1980) as the 'significant expansion of engineering education in all regions at both graduate and technician level'. Within this area 'Electronics and Computer Technology' is singled out for particular attention. For a review of the 'manpower planning' projections which underlie this expansion, see MURRAY and WICKHAM, 1982, op cit.

27. ROBERTS et al, op cit, p.258.

Training for new technology

Sheila Rothwell and David Davidson

INTRODUCTION TO THE RESEARCH

The introduction of computerised technology appears to have dramatic implications for many aspects of industry in the service sector and also in the manufacturing sector where technological change is hardly a novelty. Apart from forecasts of drastic decreases in numbers employed very little seemed to be known about the likely implications for other aspects of organisations' employment policies - the training and utilisation of remaining employees, payment systems, job design and management and supervisory structures for example - or what was actually happening in practice. The Henley Centre for Employment Policy Studies, supported by the Manpower Services Commission, therefore, in 1981, began a 2-year programme of case-study based research to see what changes were taking place in a number of different organisations(1).

All the organisations studied were at different stages of implementation - from early planning, through to completion and on to preparation for the next instalment. This is likely to be a feature of information technology since the potential applications are so widespread that completion of one tends to lead to another, to obtain greater utilisation of the benefits. Moreover the novelty and difficulty of computerisation on such a scale tends to mean that organisations usually prefer to handle it in instalments and then predicted time-scales are often doubled. Comparison and generalisation from a series of case-studies of this nature is therefore difficult but it is possible to highlight significant features or problems which are common to more than one or are instanced in other similar studies.

CASE STUDIES

Altogether over twenty applications of information technology were studied. Thirteen of them were in manufacturing industry - mainly engineering, pharmaceutical, and food, drink and tobacco, while the service industry cases included distribution, insurance, and a public utility. Five of the manufacturing cases however involved automation of manufacturing processes through the introduction of microprocessor controlled equipment. Six applications studied involved materials and production planning and control, while two related to quantity control systems. The customer service ordering and distribution system of one manufacturer (Photo Products) was also studied - the other cases were mainly in the distribution industry (see Table 1).

Table 1: Manufacturing case studies

New Technology	Company	Industry	Parent Company	Region	Employees at Establishment
Production Process	Confections	Food, Drink and Tobacco	UK	South West	1,800
	Engines	Engineering	USA	Scotland Midlands	1,200 670
	Heaters	Engineering	UK	Wales	450
	Packet Food	Food, Drink and Tobacco	UK	South East	75
	Print	Engineering	UK	South West	200
Quantity Control	Liquid Food	Food, Drink and Tobacco	UK	Wales South West	70 330
	Tablets	Chemicals and Pharmaceuticals	UK	Midlands	300
Materials Management	Allergens	Chemicals and Pharmaceuticals	Germany	Wales	450
	Drugs	Chemicals and Pharmaceuticals	France	South East	2,850
	Electronics	Engineering	USA	South East	590
	Infusions	Chemicals and Pharmaceuticals	USA	South East	900
	Photo Products	Chemicals and Pharmaceuticals	USA	South East Midlands	160 250
	Power Tools	Engineering	USA	Northern	1,400

There are considerable differences, both physical and conceptual, between automating a production process and replacing a paper based information system with a data base on a computer, yet planning methods, and organisational approaches to implementation and to handling the employee impact may be very similar. Moreover, some of the systems studied do not fall easily into the above dichotomy - automated packing and warehousing systems, like the quantity control cases, are both part of the 'production' process and the 'information' system. A materials management system may begin by replacing paper based stock records or customer accounts, warehouse supply and distribution, but also the whole movement of materials and work in progress throughout the factory, to the extent that it is 'driving' the production system.

Case-study unit boundaries therefore varied according to the stage of progress of implementation and the issues pursued. In most cases it was not the whole organisation, although in some the 'ripple' effects of change were very widespread. Thus, even if the number of employees directly affected was usually limited, the indirect repercussions were considerable, from senior management to shop floor, from sales and marketing through design engineering, purchasing and stores, production, distribution services and accounting. Almost no activities were unaffected. The research focussed on personnel policies, but the semi-structured interviews covered all organisational aspects of the change, including the reasons for innovation. These tended to be aimed at improving customer-service and/or increasing productivity and profitability; the reduction of labour costs was stated to be a major aspect of this.

PERSONNEL POLICIES AND TRAINING INFLUENCES

Personnel policies are not an end in themselves, nor are they generally seen as such by personnel directors. They are a means to an end in organisational terms: a means to ensuring that the right people are available at the right time to achieve the better output or service required. Nevertheless the importance of people to the organisation and the extent to which human resources are seen as worth developing, whether for intrinsic or extrinsic reasons, will vary from one company to another - even in the traditionally labour intensive companies we studied. 'Human capital' theory might suggest that this would depend on whether a 'labour supply' or 'investment ' interpretation of training is appropriate(2). But, as economists complain, the formal empirical analysis of investment in training by employers is beset by difficulties which relate to the sheer complexity of training arrangements, their intimate links with production activity (making the attribution of costs very difficult) and the lack of basic data.

While 'training' is only one aspect of a company's personnel policies, it is a highly relevant aspect at the present time, if only because of recruitment 'freezes' in many instances. Many contradictions are apparent, with government policies emphasising and supporting it on the one hand, but cutting Industrial Training Boards on the other; companies reducing their training departments on the one hand, but senior managers speaking and writing about the importance of training for the future in every journal and at every conference on the other. As Lindley comments, 'both government and industry appear to be more than usually convinced that they cannot afford to do much, at the same time as being more than usually convinced that much deserves to be done'(3).

Organisational decisions to go ahead with 'new technology' investment were usually found to have originated in the enthusiasm of the managing director or another senior director - often Marketing - whose support and impetus was often significant in driving through the implementation. If he had a clear philosophy of the importance of 'people' in this, it was usually reflected in the relevant team appointments and the attitude to training; as in Electronics, Allergens, Heating, Engines (in a US parent company's philosophy) and Print. If top level interest was almost exclusively 'finance' or 'results' oriented then much would depend on the commitment, power and credibility of a particular manager, either on a senior inter-departmental Steering Committee, or, more likely, with responsibility for a smaller implementation team. Thus, in Infusions the influence of the Administration Manager was significant, in Power Tools it was the Systems Manager. In Photo Products it was rather more diffused between line and systems managers, as in Confections, another company with a strong 'paternalist' tradition, and also (like most of our other companies, at least until quite recently) in a strong market position.

The existence of a powerful trade union or another similar 'interest' group might also be hypothesised to influence the extent to which 'change' decisions had a significant people and/or a training commitment built into the implementation strategy. This was not found to be there so in our cases, although companies with 'sensitive' industrial relations are generally reluctant to allow research access, or to reveal decisions dictated by union initiatives or pressures.

'Technical' influences might be expected to be the other major factors affecting the importance attached to training in the implementation process. This could work in a positive direction in that if technical skills were needed, technical training was likely to be taken seriously (as in Tablets or Confections), on the other hand if the application of information technology was seen primarily as a technical systems issue then 'training' and 'people' issues tend to be ascribed low priority. To some extent this could be seen in all the organisations studied, at some stages inevitably, or on the part of some managers, but there was also an element of technical 'luck' involved: where the new plant, the hardware or the software worked then schedules were met (approximately) and training plans maintained. Where there were serious technical delays or breakdowns then training disappeared regardless of planned proposals (Infusions, Confections).

APPROACHES TO TRAINING AND COMMUNICATIONS

Definitions of training vary considerably, at one extreme (especially in the engineering industry) some people still equate the term with time-served apprenticeship training, although this is less common in the 'new technology' context. Nevertheless even in that context some associate the term 'training' with 'going back to school' atmosphere and techniques: for example, employees, at one Liquid Food establishment and at Infusions, who were critical of the 'inadequate' training, said that on-the-job guidance and practice had been very valuable. At the other end of the spectrum, 'training' may be used interchangably with 'education','development', or even 'appreciation', 'communication' and 'propaganda'.

This wider definition is gradually becoming more generally used by managers who have obtained experience of implementing applications of information technology and have realised that concentration on formal 'training' methods and 'technical' content is too narrow to meet the new needs. The 'who', 'when' and 'what' of the training approaches found all relate to this underlying definition.

Identification of training needs could therefore be a central part of the whole strategy of implementation. Where 'communications' and 'awareness' was given high priority (as in Electronics, Power Tools and Confections) then management responsibility for it was clearly defined and an effort made to involve all the workforce through video and discussion (Confections, Infusions), or general talks and specialist seminars (Engines, Power Tools, Electronics, Heaters). Three companies (Electronics, Power Tools, Allergens) paid particular attention to their managers' awareness and understanding, even if they were not directly involved initially, since management commitment and understanding was realised to be critical in utilising and developing the benefits from the new systems and motivating employees through the many snags likely to arise in implementation. The need to break down interdepartmental barriers and resentments - particularly of 'management services' were seen as vital.

Those companies that took a broad approach to communicating and training were therefore more likely to involve, or at least consult, early in the design stage, those employees who would be affected. Photo Products, for example, developed something of an 'education-through-involvement' approach at its 'greenfield' factory, where implementation was carefully phased, and only small groups were affected by each stage initially. Packet Food's company task force decided on training needs at a meeting a year prior to the planned project completion date, but then everything was delayed by head office for fear of industrial relations repercussions elsewhere.

EXTENT OF TRAINING

The amount of time spent on 'training' in the broadest sense was thus not easy to measure since it varied according to need, to method of implementation chosen, to company philosophy and to technical delays. Managers who initially had a broad 'people-orientated' approach at the beginning, once technical delays occured and schedules slipped, were just as urgently pressing trainers to provide quick and easy methods as those who had thought little about the issues.

The general pattern was for there to be some preliminary overview and presentations made to employees at various levels at a fairly early stage - even a year or so prior to implementation; occasionally some training for systems staff; more specialised user training where needed three months or less before implementation; then guided learning and practice on 'D-Day' and from 1-12 weeks thereafter. Greater complexity of planning and skills tend to be involved in a production line application than an information system one. On the other hand, two companies, with mainly female clerical employees in the establishments affected, took the explicit view that the new systems should be as 'idiot proof' as possible and training was mainly confined to operating skills once the system came live.

SKILLS

Identification of training needs depended not only on broad or narrow definitions of 'training' but also on concepts of 'skill'. Some companies saw this purely in 'job' or technical terms - such as keyboard operating - while others included 'attitudes' such as 'discipline' in avoiding and rectifying mistakes, and 'understanding' of the potentials of the system, or the implications for others' jobs.

Most people interviewed found it difficult to talk in terms of skills. Instead they tended to talk about a variety of different types of behaviour, which can be categorised(4) as follows:

Knowledge - things known
Skills - things done
Attitudes - things felt
Appreciation - things valued
Understanding - things comprehended

It seems obvious that bringing about changes in these aspects of behaviour (and they will probably be of differential importance according to particular changes) takes more than a few days training. This is more closely related to the definition given by the Department of Employment: 'The systematic development of the attitude/knowledge/skill behaviour pattern required by an individual in order to perform adequately a given task or job(5).

Several writers(6) on managers' uses of the term 'skill' for training or recruitment for example, have shown that while it may initially appear to be a fairly objective technical concept relating to formal educational qualifications and/or specific abilities to do certain things, in practice it is often highly subjective relating to the person more than the job and subsumes a variety of attitudes which can only be summarised as 'the sort of person we need', 'the right type', 'someone who fits in'. Thus, although the definitions of 'the sort of person we need' may change in certain critical dimensions in response to the demands of the new technology, it may not change in all respects. It is not therefore easy to evaluate the jobs as becoming 'more ' or 'less' skilled.

People may acquire a wider range of less specialised skills ('horizontal' skill aquisition), and/or more or fewer more specialised skills ('vertical' skill aquisition) (see Figure 1).

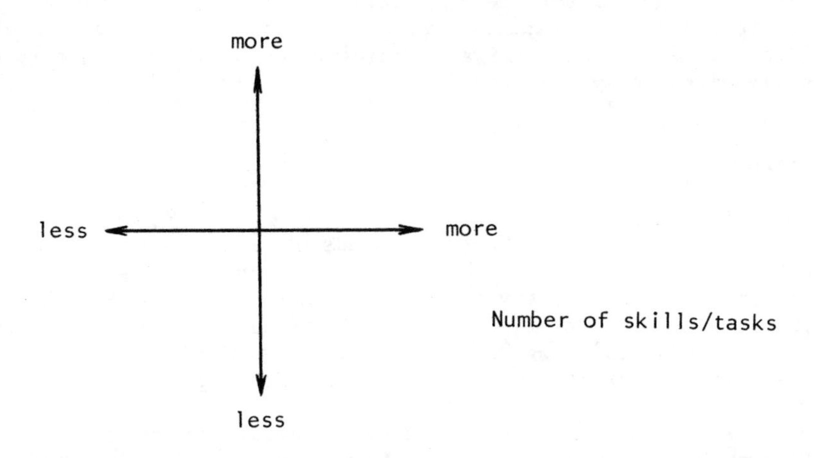

Figure 1

Maintenance craftsmen at Confections may need to acquire a wider mix of 'engineering' and 'electrical' skills (horizontal) plus new electronic technician skills (vertical) for example. Order clerks at Photo Products may need both keyboard and telephone skills (horizontal) which they have to exercise simultaneously, a much more complex achievement (vertical). Fork lift truck drivers may input data as well as shift goods around at Photo Products, while at another distribution warehouse they merely receive a computer schedule instead of a handwritten document or verbal instruction, and may become less knowledgeable through missing the human interaction and organisation gossip.

'Conversion Training' was the title most commonly given to what might most easily be understood as 'new technology' training. In 'information system' applications it usually included keyboard skills but this was seen as less important than the use and understanding of computer files and how they relate to each other, because of the importance of maintaining a highly accurate common data base. It usually consisted of some advance demonstration and practice followed by on-the-job coaching once the system went live. Sometimes it was followed up with some 'mixed' group training to try to assist employees' understanding of the relationship of their work (or mistakes) to other sections.

Management services and computer departments generally preferred to recruit their skills ready-made or else to rely on sending staff to suppliers training courses bought in as part of the software and/or hardware package - as at Electronics. Some companies chose to subcontract these activities but still found themselves critically affected when the subcontractor's key systems analysts on that project left their jobs unexpectedly.

SUPERVISORY SKILLS

Most companies, whether or not they had anticipated it, were finding that supervisory roles and requirements were changing. If these roles were already highly ambiguous the change required was not necessarily immediately obvious since supervisors were already expected to perform a wide range of duties from people management to doing the job itself(7). In practice supervisors had usually worked out their own configuration, matching what was really needed against their own skills and abilities to cope. With the automation of manufacturing processes or with materials management systems they usually had to acquire not only the ability to cope with the computerised system, but also often new team-building and managerial duties. Where they were required to ensure production or output was maintained in the transition stage, then their training was sometimes postponed. Their traditional 'fix it' skills and methods of short-cutting the formal system often had to be abandoned when the new emphasis was on disciplined routines of recording all 'inputs' and 'outputs' and adhering to computer planned schedules.

Computer and technical skills were particularly needed by supervisors at Heaters, Print, Confections and Packet Food where they had a significant role in setting up the new system. People management, group work, and team-building was more important where the operators had more direct responsibilities - as at Engines and Photo Products. Some training was provided in this but supervisors felt that more was needed to be able to move away from traditional styles to a non-directive approach. This was particularly difficult at the latter where they still retained accountability for output. Further training in financial skills and in report writing were also felt to be needed by these and other supervisors where they were given new responsibilities for cost control and meeting financial targets. The use of computer printouts to generate management information and assist in problem-solving was found to be difficult and avoided by them. In some 'office' areas of Liquid Foods and of Photo Products supervisory training was given little attention - in general women supervisors seemed to be regarded as insignificant and were merely relied on to keep the job going through the transition.

TRAINING EVALUATION AND DEVELOPMENT

Most companies realised with hindsight that not enough time had been allowed for training and that the development of real understanding takes a much longer time to develop than the acquisition of ability to input 'goods received' data, or even to operate process controlled assembly lines. The change needed to obtain the right balance between the traditional manufacturing emphasis on 'output' - getting it done and through the door - and the new disciplines needed to follow the procedures and record materials movement as required by computerised systems is considerable. As systems develop and become more integrated, so the same person may be performing an increasing variety of tasks through the system and need to build up these 'skills' and wider comprehension at the same time.

Continuing modular training and 'testing' of standards reached is being developed in a few cases - such as Engines and Photo Products.In Photo Products, the training process, combined with continuous on-the-job experience, now follows a 3½ year schedule. Even where no such complete training package was developed, an increasing number of companies were giving more attention to maintenance of individual training records and logging of 'skill' acquisitions.

TRAINING METHODS

The development of detailed manuals which could be used for training purposes in a variety of different ways (one-off or modular) and to assist in problem-solving diagnosis on the job, on a continuing (and up-dated) basis was a particular feature of this 'new technology' training which we found in many cases - although perhaps more particularly where manufacturing processes were being automated. Thus details of each operation, the processes involved, the specification to be met, the health and safety requirements, the likely faults and methods of dealing with them were all set out in words, figures and diagrams. By covering each operation in detail, rather than each job, a variety of task configuration was made possible, and scope for each operator to rotate and progress to mastering the full range of operations was facilitated in factories, warehouses or offices. Manuals were found to be particularly useful in Print, where three-shift working was introduced for the new process, but there was no supervision on the nightshift so that operators needed to be able to diagnose and deal with faults on their own.

In most cases the usual mixtures of classroom teaching and on-the-job instruction took place, but with a tendency to emphasise practice first and problem-solving through the system, with follow-up later to explain more of the 'theory' and develop understanding of what was happening, such as effects on other sections of keying-in errors. In Electronics each departmental manager gave talks to explain his work to other departments.

More use was gradually being made of other media, for example, video games, the provision of an 'old' machine, or a lunchtime 'Apple' club, for familiarisation and to 'take away the mystique'. Increasing use of company-made video cassettes for general information and communication as well as occasional specific training was also found. Limited development of simulated programmes for training was found, although there was talk of doing much more of this - much seemed to depend on the availability of equipment and the scheduling of implementation. More 'interactive' computer training programmes were felt to be needed.

SELECTION FOR TRAINING

Selection for training or for work with the 'new' technology does not seem to have been a priority issue in most of our cases. In a few, 'natural wastage' and 'early retirement' may have included the departure of those who did not feel they could adapt to change; in others only part of the job was different in a warehouse or office, or the extra training needed seemed minimal.

Considerable thought was however explicitly given to the issue at Confections where a previous attempt at implementation had failed, largely for technical reasons, but partly because training had been inadequate and selection of operators had been to a great extent on the basis of union insistence on seniority. The project team discussed whether trainability testing or special attributes would be needed for selection criteria for operators but (under the guidance of the personnel manager) this was decided against (particularly as it was not to be done at higher levels). Eventually it was agreed that selection should be jointly agreed between unions, management and the individual concerned on criteria of willingness, suitability and successful completion of training. A similar method was evolved by the work-group at Engines - to deal with what was acknowledged to be overmanning. Management saw this as a big problem, but the team agreed a straight 'last in, first out' system with the surplus men (four) to be deployed elsewhere.

The ability of warehousemen to be trained on visual display units was 'assumed' at Photo Products, and carefully handled (plus a lot of personal contact and jollying along by the Systems Manager), whereas elsewhere arrangements were made on the opposite assumption (or because union problems were anticipated) - clerks or supervisors being trained for these functions. Both Packet Foods and Photo Products were on new sites and involved mainly recently recruited labour.

Although most companies had frozen their recruitment, where new clerks were needed for data processing, schoolgirls were widely preferred because they were said to come in without preconceived ideas and to learn the techniques very easily and quickly. Otherwise, little mention was made of 'age' as a significant factor in 'trainability' - more often surprise was expressed at the case with which some older people were able to adapt.

PAYMENT FOR TRAINING

Trade Unions tend to raise this issue in some form as part of the price they try to negotiate for 'co-operating' with the introduction of new technology, or at least of receiving a share of the benefits. Managers tend to resist the concept of 'paying for training' even more strongly than 'paying for change', although they may be slightly more sympathetic to claims for a share in the higher productivity. With the growth of unemployment, however, management has found it easier to avoid claims for increases and emphasise the value of having a job at all -

plus the fact that the work is likely to be 'easier' and 'cleaner' than in the past.

Managers were more interested in seeing some concept of skill progression and linkage of pay with achievement. Modular training systems therefore were sometimes found to have been linked with (small) stepped pay increases and sometimes with new job evaluations or gradings (Engines, Confections, Photo Products). In the case of a new warehouse (Packet Foods) employees' take-home pay was less than that of comparable workers at other sites because the site provided an opportunity to reorganise working hours and to avoid overtime. Changed shift patterns also changed payments at Confections.

In the most strongly unionised cases (Confections and Photo Products) industrial relations problems centering primarily around new jobs and pay grades (of craftsmen, and of packers, material controllers, and assemblers) were experienced. Nevertheless, management objectives appear to have been achieved, with little concession, in view of the pressures on them to control labour costs and avoid repercussive effects. In the customer order section of one, however, more essential adjustments had to be made eventually and ill feelings rankled long after regradings had been negotiated, with even some management disaffection.

Problems tended to arise because of the difficulty of fitting training in to normal working hours on top of keeping the job running. Thus at one warehouse where training took place before or after normal shifts, overtime rates were paid, but in the office section a special supplement was paid, based on a calculation of the additional hours needed for training on top of the normal working hours, at an average overtime rate. Hardly any overtime was ever actually worked although extra effort and some lunchtime work was involved.

Several of our cases were non-unionised, but no major differences were observed in the ways the issues were handled. While managers there tended to stress the advantages of getting things done more quickly, they were in some respects more willing to be open in their consultations. Where union problems were feared, managers tended to say little and while this might reinforce an atmosphere of distrust it could be successful in the shortrun. Trade union officials and convenors may not know what is happening and shop stewards may not be equipped to cope with staged implementation and just accept it as part of management's traditional control over work, job design and training issues. Their members were more likely to be anxious about coping with the new technology and concerned to have a job at all.

Only one case was found of an incipient inter-union dispute, where the white collar union was claiming a monopoly on the right to input data wherever it took place, in office, factory, or warehouse. There seemed little chance of success, however in the face of determind management opposition.

MANPOWER UTILISATION

Trade unions thus tended to be more involved in negotiating to prevent - or get a good deal - for redundancy than in having a say in the details of training or job-design. Redundancies were planned in many cases but mostly dealt with through natural wastage or early retirement. At Confections they were deliberately delayed in response to union pressures and the problems of the local labour market. At Power Tools, Heaters, Electronics, and Photo Products office they had largely preceded the changes as part of a general reorganisation as well as preparation for technological change. Most companies were, however, likely to need further redundancies, probably compulsory, a year or two after implementation, even if they tended to find they needed more people than expected to maintain output during implementation.

Redeployment and retraining of people to do their old job in a new way or to do a new job in the same company was usually agreed as part of 'redundancy avoidance' procedure, but the former was more commonly found. Hardly a single instance was found of redeployment into a management services department, although there was some recruitment into it. There seemed few instances in which 'job design' or 'work flow design' and people allocation was reconceptualised: existing tasks and numbers were modified 'just enough' to fit the new technical requirements in a particular section, depending on the approach of the manager involved.

In general there was a preference for changing as little as possible, since it was recognised that change tends to provoke opposition. While it might be accepted theoretically that the four factors of 'task', 'technology', 'skill' and 'training' could be juggled and repackaged in different ways they do not seem to have been worked out in practice. Most approaches to training tended to be 'instrumental' rather than 'normative'(8). Sometimes there seems to have been a specific 'increase in job satisfaction' aim and marginal awareness of the oft analysed attributes of a good job, such as variety, autonomy, feedback, progression, but any achievements tend to have been post-hoc rationalisations. Thus people might be required to do more than one task, most of the tasks, or all the tasks in different work groups, and for different reasons. Sometimes the pattern was 'training' driven in that the necessity of incremental training for different tasks in itself encouraged more flexible work group practices; sometimes a management decision to encourage flexibility and rotate tasks led to a 'modular' training method; or sometimes any flexibility found was a result of personal or small group preference within the scope (however limited) of a particular system design. Issues of craft/technician flexibility however tended to bear more relation to pay and industrial relations priorities than to technical needs or managerial planning(9). In many instances managerial policies were also ad hoc and reactive, rather than planned strategies for change.

CONCLUSIONS

Training for new technology was thus generally ascribed low priority in practice because of the urgent technical and operational pressures experienced by organisations unused to coping with such a complexity of changes. While with hindsight they might well say, 'We should have allowed more time for training', whether they would do so in future remained suspect, since it was the intangibles of 'understanding', 'awareness', and new patterns of thinking, relating and problem-solving that needed time to be developed rather than the instant skills of doing and fixing for which a 'quick and cheap' training solution might more readily be devised. These less tangible aspects of training - which relate to the broader definitions of it in terms of education, communication, and appreciation - are the very ones demanding a more strategic conceptual approach to diagnosing needs and relating them to business developments. They may well imply a new approach to the development of human resources and organisational careers which goes well beyond devising a plan for training clerks to handle computer 'conversations', or giving managers keyboard skills. In those narrow terms, training is readily seen as a quickly applicable 'lubricant' to ease the stickiest aspects of change, but one which otherwise seems unnecessarily expensive in terms of time and money. Investment in human capital in practice seems to bear little relationship to investment in technical capital and yet investment in the former may be necessary for investment in the latter to be fully realised.

While some features of the new technology may enable organisations to do without people or rely on the market to provide those residual skills that are needed, there are many other respects in which understanding of the technology and of the business are likely to be essential and which demand investment in career development. To a large extent the balance, and approach, are matters of organisational choice, not of technological necessity, and need to be made explicit. The comment of one of our managers, 'If you think education is costly, try ignorance' has implications for those responsible for employee development and for education and training policies at both organisational and national level.

REFERENCES

1. ROTHWELL, S. and DAVIDSON, D. New Technology and Manpower Utilisation. Employment Gazette, June 1982, pp. 252-4; July 1982, pp. 280-3.

2. LINDLEY, R.M., Occupational Choice and Investment in Human Capital. in CREEDY, J. and THOMAS, B. (eds.), The Economics of Labour. London: Butterworth, 1982.

3. LINDLEY, op cit, p. 121.

4. KNOWLES, M.S. The Adult Learner: A Neglected Species. Houston: Gulf Publishing, 1973.

5. DEPARTMENT OF EMPLOYMENT. Glossary of Training Terms. London: HMSO, 1971; see also DOWNS, S. Industrial Training. in: WILLIAMS, A.P.O. (ed), Using Personnel Research. Aldershot: Gower, 1983.

6. OLIVER, J.M. and TURTON, J.R. Is There a Shortage of Skilled Labour? British Journal of Industrial Relations, 20, 1982; WOOD, S. (ed.) The Degradation of Work? London: Hutchinson, 1982.

7. THURLEY, K. and WIRDENIUS, H. Supervision: A Reappraisal. Heinemann, 1973.

8. TIPTON, B. The Quality of Training and the Design of Work. Industrial Relations Journal, 13(1), Spring, 1982.

9. For comparisions of British and German practice, see SORGE, A., HARTMANN, G., WARNER, M. and NICHOLAS, I. Microelectronics and Manpower in Manufacturing. Aldershot: Gower, 1983.

New technology in print: men's work and women's chances

Cynthia Cockburn

Not very long ago I stood at the composing room door in a well-known Fleet Street newspaper office. I looked down a line of men setting the type for the next day's edition. To someone who has worked as a typist, two things were noticeable. The first was that these men, with their rather muscular shoulders and a sort of work-shop air about them, looked somehow out of place in this office environment, perched on their typing stools, tucking their elbows well in to fit the narrow span of the typewriter keyboard. The second was that the majority of them were typing rather slowly, some with two fingers (or at least not with all ten) and looking hard at their hands as they did so. In other words, they looked like amateur typists. These were former hot metal compositors, newly re-trained for work on photocomposition. They were earning well over £200 a week, more likely twice that figure. Yet they did not look happy in their work. This snapshot picture hints at what information technology has done to newspaper composing rooms.

It has been a complex and stressful change, technically, politically and emotionally. In this chapter I want to rehearse some of its implications and try to draw some strategic conclusions from it for trade unionism. The research I have just completed on the introduction of photocomposition in the London newspaper industry was based on four case studies of newspaper firms, two in Fleet Street and two in newspaper houses on the outer edges of London. It would be possible to stay well within the parameters set by other contributions to this book and have much to report. There have been important implications in the new technology for organisation and control of work; for pay and hours; for shift patterns; for levels of productivity. But I want to tease out of the story some other factors - more immediately social and subjective. I want to look at what such changes have meant for the men involved, the feelings they express about them, the meaning they make of them.

But first I should say a little of what the technology and labour process were and what they have become. Composing type for letterpress printing was a process that used to involve molten lead alloy. It had two important parts to it. The first was type setting. In newspapers this was done on a linotype machine: quite a substantial machine about 7 ft high and perhaps 6 ft across, with many moving parts. The compositor (or 'comp') sat at a keyboard, tapping characters. Each tap released a little brass matrix, which fell down a chute to assemble with others in a line of text. When the line was tight, properly spaced, the operator pulled a lever and molten metal was pumped into the type faces to produce a solid, hot 'slug' of type about an inch high and several inches long. The matrices were subsequently recirculated for further use and after the printing had been completed the lead itself was recycled back into the melting pot.

The second main process was imposing all these slugs on a flat surface known as the 'stone', in the right reading order, with illustration blocks and spacing pieces, to make a newspaper page. The whole, often hundreds of parts, was locked up tight in a steel frame or 'chase' and a mould taken of it, which was subsequently sent off to the foundry for casting of stereo plates to place on the cylinders of the printing press.

The significance of this labour process, for the craftsmen who engaged in it, was rich. It was an esoteric occupation: an outsider would have had little idea how to do it, would not have known the use of tools, the protocols. For instance, the keyboard of the linotype had 90 keys; it was quite different in lay from that of the typewriter. Only a practised linotype operator had facility on it.

Again, imposition involved a special system of measurement, using not inches or centimetres, but picas and points. The job called for deftness, knack and a degree of heft. There was a good deal of jargon in current use. Comps felt, and often behaved, like the priests of a mysterious religion. The controlled production to considerably greater effect, sometimes, than their managers, because when it came down to it, only they know actually how to get the work done. If they did not co-operate, with the employer and with each other, the newspaper would not reach the news stands and many thousands of pounds could be lost overnight.

Unlike the new composing room, this was a factory-type environment. It was hot, dirty, noisy. Those things might seem unpleasant in a workplace, and indeed they were sometimes complained of. But the curious thing is that such conditions combine with the presence of machinery and metallurgy to make something that men find highly desirable: a manly job.

The only way to take part in any of the phases of this labour process was to have done an apprenticeship in the whole compositor's craft. It used to be a seven year stint, starting when the lad left school. Adult trainees were rare and were frowned upon. Composing for print was an effective pre-entry closed shop, especially in the news trade. Comps and their trade societies, today their trade unions, have worked diligently to develop and maintain this closed shop over hundreds of years.

So, finally, there was a distinct status attached to the job. The compositors compared themselves very favourably with unskilled men. In fact they sometimes reffered to themselves as being among 'the gentlemen of labour'. It is an unfortunate fact, but many comps have hard words to say about the less skilled workforce in print. They pride themselves on the high pay they can earn and scorn the 'Natty' (derived from NATSOPA, the National Society of Operative Printers, Graphical and Media Personnel, a union of less-skilled print workers) for lack of intelligence, lack of loyalty to the trade, lower earnings and for his union's lesser clout.

It is useful to recognise that this self-representation by the skilled man had a bearing on life and relationships not just in work but also outside. In the 19th century and even to the present time, print craftsmen have been proud to be heads of household, breadwinners, able as a rule to keep a wife and children on their own pay. They were proud to pay sufficiently high subscriptions to the trade society to make it an effective resource for the family in hard times. The comp was content that his craft union was characteristically male. It was another sign of the impotence of the unskilled that they had to share their unions with women.

Women were seldom if ever apprenticed to the craft in the 19th century and those that slipped into the occupation did so as part-comps, in 'unfair' houses, usually setting type only. And they met concerted opposition from the male craft unions. There were several vigorous battles over the employers' hankering after cheap female labour.

So far, there is nothing unusual in this account of a craft occupation. What has been distinctive about composition is that the craft has survived until the present day. It was, in fact, quite unchallenged until well after the second world war, by which time most 19th century crafts had been replaced by modern industrial methods. Comps remained a high-paid and a costly group of workers, their 'chapels' or trade union branches retained an apparently unshakeable hold on the print production process. In the boom conditions of the fifties and sixties, this frustrated the newspaper owners to the point of desperation. The comps were continuing to keep labour in short supply and holding down output at a time when demand for advertising and print generally was growing. The employers badly needed a weapon to free up the labour market and break the craft 'stranglehold' on the composing room.

The new technology that has served this purpose was a postwar development in the USA, which rather slowly drifted across the Atlantic during the sixties and seventies. It is computerised photocomposition.

What is this new system? At its simplest, what the men call an 'idiot' system, it is no more than a series of simple keyboards, typewriter-style, on which the 'tapper' sets the text. Instead of producing a page of typing, the keyboard produces punched tape. This tape may then be used to drive a photosetting machine. This piece of equipment contains a photonegative of the letters of the alphabet, lenses to change the size, and a source of light that flashes in order to print a tiny photograph of each letter required, one by one onto a strip of white bromide paper. The photosetter, even in such simple systems, usually has a little computer capable of taking decisions concerning line-breaks and the hyphenation of words, formerly left to the intelligence of the compositor. The skill in 'idiot' setting is little more than typing. The skill in operating the photosetter is nil. One simply feeds in the tape at one end and receives the bromide at the other.

It will be obvious now what the 'stone hand' has become: a 'paste-up hand'. The paper print is cut up and affixed to a card, according to a lay-out prepared for him, ready to be photographed and laid down onto a printing plate. This 'lick and stick'job needs barely any training. It does need a good eye and deft clean fingerwork. This and keyboarding therefore proclaim themselves, within current social stereotypes, to be 'women's work'. It is easy to see that it was not traditional craftsmen but cheap female labour that the manufacturers of the new equipment had in mind.

These 'idiot' systems have quickly been surpassed however by more complex and efficient technology. A modern system is likely to have a video screen associated with the keyboard, so that the matter is visible as it is set and may easily be corrected, then or later, simply by over-typing. Once again, advanced photosetters are not in reality photographic at all, since the computer itself can produce as instructed, by digital techniques, a range of differently shaped and sized characters. The possibility of locating whole blocks of paragraphs and pictures together on a big screen is just now beginning. This avoids even the job of paste-up. The information shown on the large of this page-view terminal is stored on magnetic disc and can drive a photosetter capable of emitting an entire page. Soon it will be possible to set straight onto a printing plate from such a page-assembly on screen. Alternatively it will be possible to by-pass the plate altogether and print from computer to paper by inkjet - sprayed print. Meanwhile of course the electronic forms of media may in any case be making redundant some of the functions of conventional newspapers. The shape of the general printing trade too is changing, as a good deal of material that was once printed in a trade print shop is now xeroxed and offset by office workers in in-house print rooms.

It will be evident, then, that the new composing technology is much faster, more efficient and flexible than the old, whoever operates it. It is versatile. It enables the newspaper to interact with other media. Those would have been reasons enough for a newspaper owner to invest £2 or £3 million. But the over-riding reason, over and above all this, for technological innovation in the press, has been the opportunity it offered to break the grip of chapel control in print. Because the compositor could now be outflanked. Either he could be obviated by putting a less trained person into simplified and fragmented parts of his job. Or the job itself could be shelved. Because the logical use of photocomposition is 'direct entry'. The editor, the reporter or the advert typist can put their own typing into the computer memory, whence, once it is corrected, it can activate the photosetter direct.

In the USA, where the International Typographical Union had less power to resist the press owners, this does indeed now happen. There is, however, to date, only one newspaper firm in Britain that has defied the unions, using computerised composition on its own terms. Elsewhere, though the number of comps has been thinned out, by new technology and the recession, the former craftsmen have had to be taken on to work the new systems.

In effect, the National Graphical Association (NGA), riding now on those decades of closed shop and craft control, has succeeded in slowing down this process of change and forcing the employers into a sort of half-way position. For the press owners, it is a botched job. And for the comp, the cost has been to retrain for new work, a process which can be painful, especially when that work has none of the cachet of craft.

These are, many of them, men in their fifties and sixties. They are masters in the old ways, but many are somewhat set in those ways and no longer quick to learn. This accounts for the comps I saw tapping rather slowly and with such unconventional fingering. Of course they cannot afford to work too fast, for fear of working themselves out of a job. But, one way or another, they are, from the employer's point of view, something of a spanner in the works, slowing down the production process rather than enhancing its new potential.

This is a destructive experience for the craftsman too. The fact that pay has increased and hours of work have diminished as part of the deal for taking on new technology is not altogether enough to mitigate the discomfort. A craftsman likes to feel he is doing something clever and useful for his money and with his time. So for many it is a holding operation, until retirement, or until some new opportunity comes along - perhaps self-employment.

For the craft as well as the individual it is a gloomy prospect. There are sure to be many voluntary redundancies and early retirements over the coming years as the increased productivity of the new systems makes itself felt. This is specially so in recession when newspapers cannot expand their advertising pages to take up the slack. In at least two of the firms where I made case studies, the retraining of the original craftsmen proved insufficient for the employer, and within a few years loss of some composing jobs followed.

Beyond this, the loss of openings for young people will be incalculable. Thousands of school leavers who might otherwise have seen this as a 'good job', as it always used to be, will not get an apprenticeship for this skill. And the men themselves will have to accept working alongside people whom they consider 'trainees', or less than fully-apprenticed comps. The men may well be strong enough to bring newspapers to closure rather than allow female typists actually to take over the composing room. But, for instance, *Times* comps have had to see the *Supplements* hived off by Rupert Murdoch to a trade typesetting firm near Fleet Street which does employ women on the keyboards.

What about the union? The technological crisis has been calling forth some profound changes in strategy in the National Graphical Association (NGA). The first challenge of course has been to negotiate photocomposition agreements nationally and house-by-house on the best possible terms, exchanging co-operation on new technology for shorter hours and pay increments. They have also to try to ensure that in the transfer there is no loss of jobs and no deterioration in chapel control over production. One important condition, in some firms, has been that the work is not only NOT subject to further division of labour, but is in fact carried on in a way that is more integrated than heretofore. In integrated systems the retrained craftsman takes a turn at all aspects of the job, including input on the keyboard. At best, the chapel controls the rota. Such measures can reduce the extent of de-skilling and dissatisfaction with the job.

At another level, though, the NGA are having to intensify their efforts to amalgamate with other unions, where there may be demarcation problems exacerbated by the new technology or where soldiarity may be particularly crucial. Hence the NGA's amalgamation with the Society of Lithographic Artists, Designers, Engravers, and Process Workers and attempts in 1982 (so far not successful) to join forces with the National Union of Journalists. So the new technology is giving an added impetus to the theme of 'one union for the media'.

Secondly, the NGA is working on a new training scheme which will do away with time-served apprenticeship altogether and introduce a system of modular training, which the trainee could enter at any age, take at any pace and which could enable specialisation in one or more of many print occupations. This means (in spite of precautions the NGA is trying to build in) the gradual decline of the pre-entry closed shop. Instead, the emphasis will shift to winning a post-entry coverage of the industry.

Finally, the union has a new recruitment policy. The NGA until quite recently was almost solidly male in its membership and entirely male in its outlook. One comp I talked to a year of two ago actually thought that women members were not permitted. He was wrong. But he may be excused for not knowing, because the few women that did exist within the union's ranks were anomalies buried deep under a structure that was otherwise masculine. However, a transfer of engagements with a smaller 'industrial' union, the National Union of Wallcoverings, Decorative and Allied Trades, in 1979 brought into the NGA about two thousand women clerical members. The NGA itself was beginning to see advantage in protecting its craft members from employers' use of 'tele-ad typists' and secretaries to set newspapers by direct entry. It began to think of deliberately enrolling this kind of member. Besides, offset and xerox operators in the basements of banks, businesses and local authorities, to say nothing of women newly entering the new occupations of photosetting and paste-up, really were becoming the new kind of print worker. Perhaps, therefore, 'craft' should relax its boundaries to include these too. The NGA's new recruitment policy was advertised under the theme 'Let's work together'. Under that caption appeared photos of the new kind of member envisaged. Many were women.

Many of the contributions to this book have taken the perspective of the firm, when discussing new technologies. It may seem that already, in taking the perspective of the employee and the trade union, that I am presenting a transverse slice of the reality of technological innovation. It is one that I think it is vital to understand and to represent: the experience of the person that works and finds his livelihood in, or is dispossessed from, a technological labour process.

However, I want to go one step further and end by considering technological innovation from the point of view of a group who have never till now been present on the print shop floor: women. Because, seen from that perspective, new technology looks different. It looks different because rather than destroying an occupation it opens up new ones for them. It also looks different because it represents a chance to remodel trade unionism better to serve women's interests.

Let's go back a little. The skilled craftsman has been the most successful instance of trade unionism, successful in terms of sustaining organisation, winning good pay and conditions from the employers, controlling the labour market. They did it through a sustained struggle. It has to be said however that it was achieved by climbing on the backs of other groups in society, in particular the less skilled men - and women. The chosen craft strategy against the employers was always one of exclusivism, defining skilled men as a distinct stratum with interests that set them apart from the rest of the working class. Of course, as we've seen, the craftsman's relative strength, too, was predicated on women's domestic work and sexual subordination. Secondly, their technical knowledge was not put at the disposal of the working class and of women but sequestered as private property. History makes it clear how and why this happened, but we are free to look for an alternative philosophy today, recognising the disadvantage that has resulted for women.

Continually, when challenges arose that could have been resolved in class solidarity (that is, in the spirit of an egalitarianism that would have included women and male labourers in the concept of the working class) they were resolved by the skilled men instead by recourse to patriarchal values, to their identity as men. As men they gained their ascendancy in the sex/gender system by subordinating women and some other categories of men. Men's masculine identity has always, in this way, been in danger of being in contradiction with their class identity and interests. It has been a disastrous historical process - for women, but also, in the long run, for men.

Now look what is happening today. The new technology is tending to wipe out distinctions of men's work and women's work. It is opening up a few situations to women. This is happening at a time when women are asserting their own definition of themselves in the women's movement, their capability and rights, their sexuality. There is no doubt at all that the men in my study, many of them, felt in a pincer movement between what the employers were doing to their manly craft and what women were saying and doing. Of course some men, a few, who dislike what power has done to men, are pleased to see the promise of a little shift in the balance of that power. But many felt anxious at the prospect of being seen as 'glorified typists' in their new style work and of having to do work that 'even a woman' could now be seen to do. Others just felt angry.

It is not, of course, only the line between men's and women's work that is being redrawn. The new technology is taking all the old work definitions and throwing them around. It disturbs the hierarchical relations between mental and manual work, craftsmen and technicians, journalist and printer. And this is what one might expect, since new technology in print is just part of a restructuring of the working class that is being achieved, or attempted, on a much wider scale by strategies of capital and policies of the state.

It is a moment of peril for trade unions, for working class organisation generally. But perhaps it can be turned to good account. The door is open for a new set of expections we can have and demands we can make about trade unionism, and a new understanding of skill. We can work to make trade unions reflect women's needs as much as they now reflect men's needs. That is a very wide range of changes. In this case let us limit consideration to one: the unions could become educators of a broad and social kind, in place of masonic lodges for a privileged few. They can become places where women learn confidence, encourage each other and are encouraged by men to train for skilled technical work, and where together we can develop a social critique of technology in place of an unquestioning identification with it.

I just want to end with a few thoughts on the question of skill and deskilling. Is new technology in composition destroying skill? The simple answer has to be yes, even when, as in the stronger chapels, it has been possible to re-integrate all parts of the composing job and retain them in the hands of ex-craftsmen. But in a way the question itself is misleading. It is necessary to distinguish at least three facets of skill. There is skill in the person: what I am competent to do. There is skill in the job: what the job demands of me. And there skill, finally, as defined for political and economic purposes: what a union says the particular expertise of its members is.

This accounts for the contradiction that one can gain and lose skill simultaneously. Skill in the person is increasing with new technology. The men often say, learning new technology, 'it's another string to my bow.' Skill in the person accumulates over the years. You can never take it away, though parts of it may get rusty. But skill in the job, that can go 'right out the window' as the men put it. It may be negligible compared with the worker's overall competence. What is more, it can come completely out of kilter with the political definition of skill. In which case the union has to abandon its defence - and this in a way the NGA is doing. And in so doing it is hurting many of its disbelieving old members.

There are two further points worth bearing in mind. One is that the definition of skill is political in a second manner also. When a job is defined as skilled, it is always a man's job. Very little of women's work in industry is defined as skilled. This is partly because women have been excluded from those male occupations, but partly also because men have used their power as a sex to build up the prestige of men's work, and women have not been able to do this. Their occupations have remained under-valued just because they are associated with women in a world where women are not taken seriously as workers. So there are ways in which a complete shake-up in concepts of skill could greatly benefit women.

The second point is this however. Women can help to make better sense of skill. Skill ought rightly to include an evaluation of the product. This may seem a little idealistic. But the initiatives of 'workers plans' such as that of the Lucas Aerospace unions' Combine Committee, show that others are thinking in this way. Skill used to produce armaments is not skill. Skill used to produce a very fine looking pair of boots that let in the wet is not skill. Skill used to produce a newspaper at great speed night after night, completely free of spelling errors, but a newspaper that is heavy with anti-working class invective, that is no skill. Nor is it a skill that imposes, so deftly and efficiently, the photos and text of a newspaper page, getting them all trim and square, when the page in question is Page 3 of The Sun. Women, who have tended to be defined as the consumers of this world, less entranced by technological processes than men, closer to their impact on human life, may perhaps help make this much needed re-evaluation of skills.

If new technology does nothing more than jolt us into some awareness of our responsibility as printers and journalists for the content of the press it will have done something useful. Perhaps in a wider sense we could hope that new technologies of all kinds will jolt us into a more active awareness of the need for socially-informed choices over production processes and products.

AUTHOR'S NOTE

Research funded by the Social Science Research Council and carried out at City University. This paper selects themes from the book Brothers; Male Dominance and Technological Change, London: Pluto Press, 1983.

Conclusions

A history of 'new' technology

Craig Littler

> 'During the next few years, we shall probably see more and more robots, so that ultimately business and factories will be run by only a small proportion of the people now employed.'(1)

THE ANALYSIS OF TECHNOLOGY

Assessing the impact of new technologies and the time scale of development is no easy task, as the above quotation, published in 1933, demonstrates. Up to the present historians have not integrated fully the fields of economic history and the history of technology(2). Because of the neglect of the history of technology there is no generally accepted account of its development. In particular, there are serious problems in classifying the various stages of development; a primary task if we are to examine technology as a causal sequence.

A common answer to the above problem is to adopt an ad hoc classification; for example, the age of textiles, the railways age, the age of electricity, the age of the internal combustion engine and, currently, the electronics age(3). Clearly, the analytic usefulness of such ad hoc lists is largely nil. They encapsulate newspaper headlines, but little else.

Bright attempted a comprehensive analysis of mechanisation and automation. He defined 17 levels of mechanisation and suggested that there was a continual trend towards automaticity. Bright designed his 17 levels of mechanisation as an analytic tool to generate mechanisation profiles of specific production lines. As such there is no attempt to relate the various levels to historical phases, something which is made difficult by Bright's tendency to see a continuous trend such that all technical change is conflated into one dimension of relative automaticity(4).

Bell critises some of the weaknesses in Bright. Essentially Bright's levels of mechanisation are based on a detailed consideration of transformation processes, and other activities such as materials handling or maintenance are only briefly considered. A different analysis has been provided by Bell(5) and, following him, Coombs(6). The argument is that all production processes consist of three different functional activities: (a) the transformation of work-pieces; (b) the transfer of work-pieces between work stations; and (c) the co-ordination and control of (a) and (b). These three activities can be mechanised or manual. Moreover, they are not just analytic distinctions, because it is further suggested that there have been three phases of mechanisation which have been the successively dominant form over the past 100 years or so across most industries.

The first of these phases (primary mechanisation) ran from the middle of the nineteenth century to the end of the century and placed the emphasis on using power-driven decentralised machinery to accomplish transformation tasks. The second phase (secondary mechanisation) from roughly the start of World War I to the 1950's, placed the emphasis on using machinery to accomplish transfer tasks. The third phase (tertiary mechanisation) which began during the second World War and is still continuing, has placed the emphasis on using machines to achieve control functions(7).

Table 1. Engineering model of evolution of production process(8).

	Primary Mechanisation	Secondary Mechanisation	Tertiary Mechanisation
1850	beginning		
1875			
1900	spreading across sectors and maturing technically	beginning	
1925			
1950	continuing but increasingly likely to occur together with secondary or tertiary mechanisation	significant diffusion and increasing technical maturity Further diffusion restricted by product markets	beginning in some industries and slowly becoming more flexible
1975			flexibility increasing

In fact, such an argument is not sustainable at the level of generality which emcompasses all industries: instead it is possible to outline an 'engineering model' of change which refers to the metal-working industries, and then note the differences from other broad sectors. Thus Table 1 sketches such a model of development.

The shift from one predominant phase of mechanisation to another is associated with 'bottlenecks' and with diminishing returns in relation to existing paths of development. Thus, the rapid increase in the productivity of late 19th century machine tools resulted in bottlenecks over transfer systems. The secondary phase of mechanisation typified by assembly-line methods and Fordism (see below) solved some of the problems of synchronisation and production imbalances by mechanical handling technologies. These technologies created the mass production industries which, however, faced a restricted diffusion potential because of variations and fluctuations im many product markets, making dedicated automation impossibly expensive. Thus a new bottleneck arose from the problem of extending assembly-line production beyond the mass-production industries without control innovations which permitted flexibility. It is in this context that we should examine information technology.

However, the engineering model of change needs to be distinguished from other models such as that of the process, pre-planned industries, many of which combined the simultaneous development of primary, secondary and tertiary mechanisation from crude beginnings to the sophistication of present-day computerised chemical works. Different again is the textile industry model of change where developments have tended to combine primary and tertiary mechanisation resulting in banks of computer-controlled spinning machines, but often leaving transfer mechanisms as one man with a squeaking trolley.

One of the factors which has tended to fix the engineering model of change as the most influential one is the origins of the leading management movements. These (Taylorism, Fordism, etc.) have usually arisen from the practices of the engineering industry, at least since the early industrial revolution with its textile industry paradigm.

Thus, associated with each phase is a wider movement which has an ideological and social dynamic as well as a technological one:

(a) the consolidation of the factory system culminating in Taylorism;
(b) Fordism;
(c) ITism - the emergent ideology associated with information technology(9)

These broader industrial movements mediate the impact of technological changes, such that the 'design space' hypothesised by John Bessant is closed off in a particular way. Indeed over time the notion of a design space can cease to exist. But before we can understand how this process occurs, it is necessary to outline the evolution of production processes in relation to the ideologies of Taylorism, Fordism, and ITism.

THE EVOLUTION OF TECHNOLOGY AND IDEOLOGY

In a fundamental sense we can see the developmental sequence as one of phases of increasing systematisation; at first this involved increasing centralisation. The modern factory represents a concentration of capital and labour plus the application of mechanical power. However, many accounts of the industrial revolution give a misleading picture of the nature and evolution of the modern factory. A more accurate typology of factory development can be derived from the work of Chapman(10).

Chapman distinguishes between two types of traditional workshop. First, there is the 'centralised workshop', which brought together previously independent craftsmen or domestic workers under one roof. But the definitive feature is that it is a single-process workshop, such as the mediaeval fulling mill. In the 17th and 18th centuries there emerged workshops for dyeing, stocking frames, tape looms and so on, all of which represented a concentration of capital. Chapman points out that one cannot grasp the essence of these workshops in terms of the use and application of power. Some workshops used power machinery, and some (eg. tape looms, stocking frames) did not. Nor did the use of powered machinery necessarily imply a greater fixed capital investment than with the use of manually-operated machines: 'The simple mills of the period rarely generated more than ten horse-power and....seldom represented an investment of £100 to £200,' which was a sum often surpassed in another manual workshop(11).

The second phase of development was that of 'proto-factories'. This refers to workshops where there is a centralisation of several production processes, usually for reasons of enhanced employer control, and where an extensive sub-division of labour has not yet occurred. In such workshops sometimes mechanical power was used, and sometimes manual, and quite often a combination of both. The proto-factory was widespread in several industries before 1775, and the fully evolved factory made its appearance only slowly.

Chapman's typology of early production systems implies that the logic of industrialism was slow to work itself out. Thus Chapman emphasises that 'even in the cotton industry it was to take another half century before the implications of Arkwright's basic idea were fully worked out.'(12) Moreover, the pattern of development of cotton textiles was no more than the harbinger of future developments, rather than the typical form adopted by other industries. Many industries, like the Birmingham and Sheffield trades, retained a structure of small workshops and enterprises, which involved personal and diffuse relations of subordination(13). As Hannah argues, 'it was not until the later era of large machine tools, engineering standardisation and the assembly line that they were to gain access to internal economies of scale comparable to (and indeed greater than) those in the cotton industry.'(14) What this implies is that there continued to be a dispersal of labour and of industrial capital across many small intensely competitive firms such that it was not until the 1880-1914 period that we can first see the development of the fully-evolved factory system in some industries.

By this time two sets of developments had occurred: first the development of an identifiable machine-tool industry which acted as the main transmission centre for the diffusion of new methods and techniques across the entire machine-using sectors of the economy. This occurred because of what Rosenberg calls 'technological convergence', meaning that most firms faced similar problems when processing metal and utilised a relatively small number of broadly similar productive processes. Thus by the 1880's the milling machine had been perfected, replacing hand filing and chiselling; so had the turret lathe, and the universal grinding machine. In fact all these early machines were universal machines, which means that they were non-specialised and could perform quite different operations according to the speed and position of the tool. As a result, the flexibility of the machine is sought at the cost of considerable non-productive time, as the greater part of the time is spent in setting up. This, in turn, requires intelligence and skill by the machine operator.

The perfection of many of the 'stand-alone' metal-working machines by the 1880s created a bottleneck in relation to labour and the traditional social forms for the reproduction of skills. According to one engineering worker at the time:

> 'Improvements in the production of tool steel have enabled machine owners to speed up even older appliances four and five times faster, and to take cuts two and three times heavier than formerly. The development is no less swift than persistent. A machine operative in an up-to-date high speed engine shop is equivalent (as a producer) to 10 to 12 skilled craftsmen ten years ago, and to fully 50 skilled workers 30 years ago'.(15)

As a result, skilled manual work became subject not just to technological change, but to technological change on the basis of certain assumed patterns, which were provided by the second development; namely that of systematic and scientific management, otherwise known as Taylorism. Broadly Taylorism envisaged:

(1) Decomposition of jobs into the simplest constituent elements associated with the divorce of planning and doing, and the separation of 'direct' and 'indirect' labour.

(2) Managerial control over training

(3) Minimisation of skill requirements and job-learning times, leading to the maximum substitutability of workers and work teams and the potential of using cheap marginal or immigrant labour(16).

Taylorism can be seen as the ideological expression of the final consolidation of the factory system. From a different perspective, its timing was such that is associated with the transitional phase from primary to secondary mechanisation, and from competitive capitalism to the later corporate capitalism.

However, the basic design philosophy of the typical factory did not change between 1840 and the turn of the century. Taylorism was not intrinsically systemic. It was possible to re-organize one department along Taylorite lines and leave the rest of the factory operating under different methods. Equally, the average production system still consisted of a crude assemblage of unintegrated bits of machinery.

The influence of Taylorism is still the subject of debate amongst sociologists and economic historians, and certainly the British pattern of development differs from that in the USA(17). Nevertheless, the direct and indirect influence of Taylorism on factory jobs and technology design was clearly considerable. In the United States and, more slowly in Britain, Taylorism, with its underlying principles of job fragmentation, tight job boundaries, and the separation of mental and manual labour, became the predominant ideal for job design. However, in practice there are limits to the division of labour implied by Taylorism. As Adam Smith realised, the division of labour depends on the extent of the market. If a certain piece of work involved ten operations it would not be economical to employ a specialised, detailed worker for each operation if the total volume of output only required the time of one person. Thus, decomposition of tasks and Taylorite principles depend on mass markets, mass production and the velocity of throughput.

At the levels of both technology and management, there were severe problems of co-ordination; of re-integrating the new division of labour. Taylor himself advocated administrative solutions to these problems - a bureaucratisation of the shop-floor - but Henry Ford evolved technological solutions as well. Fordism is important because the production technology of automobiles established the pattern for technical change in the modern mass production industries throughout the 20th century.

The linkage of the division of labour and mass markets was realised clearly by Henry Ford. He largely established, captured and maintained a mass market for automobiles between 1908 and 1929, when the last of over 15 million model T cars rolled off the assembly line. By that date the USA had about 80% of the cars in the entire world, a ratio of 5.3 people for every car registered at a time when cars were a comparative luxury in Britain(18).

According to Meyer(19), Fordism as a model of production worked out by Ford between 1908 and 1913, had four basic elements:

(1) Standardised product design
(2) The extensive use of new machine tool technology.
(3) Flow-line production
(4) The implementation of Taylorism in relation to work processes.

As I have already suggested, the key to understanding the factory operations of Ford lies in the standard design of the Model T car. An observer in 1917 refers to a case where the Ford engineers changed the appearance of the hood and fenders, and goes on to record that 'The first month saw production curtailed by 50% and it was nearly three months before the entire organization could be geared up for the stipulated work (20). Thus, even apparently minor changes in design can cause a shutdown of many months or the replacement of all tooling whilst managers and workers re-discover the most efficient way to produce or assemble the new product.

But Ford's industrial expansion was associated with, as both effect and cause, significant changes in the design and construction of machine tools. Up to the end of the 19th century most machine tools were still general-purpose machines which relied on a repertoire of skills by the operator. But the emergent automobile industry, adding to the impact of the bicycle industry, sparked a new and intense phase in the design and specialisation of machine tools. New design features made them semi-automatic with special controls to change or reverse speeds. There was a rapid development of special jigs and fixtures which simplified the setting-up operations and controlled the tolerances of work. Essentially the machines were designed to run continuously at high speeds once they were set up. The widespread use of specialised machines in workshops was associated with a new division of labour based on a distinction between the 'set-up man', whose job consisted of starting the machines' operations and debugging it, and the semi-skilled machine minder whose only tasks were to feed the machine and remove the finished pieces. By 1914 about 15,000 machines had been installed at the new, vast Highland Park plant and company policy was to scrap machines as fast as they could be replaced by improved types.

However, the greater perfection of machine tools does not indicate the qualitative shift represented by Fordism. As Bright asks: 'How do collections of machines evolve into the production line and the line into an integrated, highly automatic sequence? What is the essential principle?'(21) The answer is that Ford perfected the flow-line principle of assembly work. This meant that instead of workers moving between tasks, the flow of parts is achieved as much as possible by machines (conveyors and transporters) such that assembly workers are tied to their work position and have no need to move about the workshop. A crucial consequence of this, is that the pace of work is controlled mechanically and not by the workers or supervisors. Thus instead of the increasing automaticity of individual machines, there is a shift towards the mechanisation of a large span of the manufacturing cycle including transfer processes.

Associated with the new fixed-speed moving assembly lines was an accelerated division of labour and short task-cycle times. Ford pushed job fragmentation to an extreme. For example in 1922, Henry Ford records a survey of jobs in his plants:

> 'The lightest jobs were again classified to discover how many of them required the use of full faculties, and we found that 670 could be filled by legless men, 2,637 by one-legged men, two by armless men, 715 by one-armed men, and ten by blind men. Therefore, out of 7,882 kinds of job....4,034 did not require full physical capacity.'(22)

Assembly lines, mass production and the elements of Fordism are now a commonplace factor in our industrial world, but in 1914 the new technology at Highland Park fascinated and overwhelmed contemporary observers. Thus, H L Arnold described:

> 'Long lines of slowly moving assemblies in progress, busy troupes of successive operators, the rapid growth of the chassis as component after component is added from overhead sources of supply, and finally the instant start into self-moving power.'(23)

THE LIMITS TO FORDISM

But Arnold, along with other observers saw a false promise of the future. Mass production methods did spread beyond the USA to penetrate the car plants of Europe; beyond automobiles into electrical products - a diffusion which was assisted by the development of multi-national corporations and the internationalisation of technology in the inter-war years. But there were strict limits to the diffusion of the 'new' technology of 1914: limits set by the nature of the market. As recently as 1969 only 25%(by value) of US industrial production was mass production, whilst 75% was still produced by means of batches(24).

In general it is still the case that the vast majority (up to 80%) of engineering components are produced in batches of less than 1,000. This is significant because traditional batch production costs between ten to thirty times more than mass production of an item. This is because of the need to continually re-set the machines and the considerable delays in the movement of components between machines. Most items spend long periods collecting dust on the factory floor queueing for the next process.

In looking at batch production in the early 1970s, Turner noted that it is complex, because succeeding batches require different machining operations in different sequences. This entails a large amount of variety, great uncertainties, and complex work-flow patterns. Associated with this complex pattern is a lack of complete knowledge of the production system by management, such that instead of a production planning programme there is a monitoring of work-in-progress by an army of progress-chasers and harassed foremen(25).

In the early 1960s one solution to these problems was the evolution of so-called 'group technology'. This originally was a technical term referring to a new lay-out of production based on grouping together all the machines necessary to complete a particular type of component. This in turn was based on classification of all components, standardisation as far as possible and grouping the components into 'families'. It improved machine utilisation and it speeded up the throughput of work by simplifying the flow of work. But whatever the managerial advantages of group technology, after a limited spread to about 10% of batch engineering firms in the early and mid-1970s, the process of diffusion came to a halt. This was because the information burden of setting up a reliable group technology production system was too great - there were too many variables and too much unpredictability.

It is these two facts - the restricted reality of mass production and the enormous information burden of automating batch production methods - which provide us with an historically-based key to understanding the impact of IT in manufacturing.

However, the first steps towards automation and tertiary mechanisation were not in the direction of flexible automation, but towards dedicated automation. As a result, despite the 1950s and 1960s debates about the consequences of automation, in the event the expected impact did not transpire and its application was largely confirmed within specific industrial sectors.

The 1950s did produce one theorist - John Diebold(26) - who foresaw the potential of automation in terms of the notion that the production process could be totally integrated, incorporating design, planning, scheduling, manufacture and transfer. This is the final step towards a total synthesis of the entire production system. But it took the development of the microprocessor in the 1970s to make the systematic integration of office and shop-floor tasks possible.

ITism

What, then, has been the impact of IT so far, especially in the British context? The preceding ten chapters point towards a number of common themes which have been outlined by Graham Winch in the introduction. First, the programmability of the new technology combined with cheap data storage, creates the potential for the first time of automating non-mass production. Given this revolution in technology, the principle of automation is spreading beyond the confines of mass production to traditional batch production areas. For other sectors such as the continuous process industries, and cars themselves, there has been no 'automation revolution', but rather a continuous evolution of technology. Nevertheless, this 'technological convergence' of mass and batch production, both areas utilising similar technology, means that one of the environmental contingencies bearing on organisations has been severed: it now becomes possible to automate even under conditions of market fluctuations and market unpredictabilities.

Second, there is no simple technological determinism. The new technology, like the 'new' technology of earlier waves of mechanisation, permits a choice and variety in relation to work organisation and job design. Similarly, the impact on skills shows no unilinear tendency. New technology does not simply represent a new phase of deskilling, the picture is more complex.

In this vein, John Bessant argues cogently for a concept of 'design space', a variable area of manoeuvre in relation to any new technology. At present we seen to be in a phase of development where the 'new' technology is genuinely new, that is it appears to be malleable and to offer a range of options - centralisation versus decentralisation; enhancement of skills versus the polarisation of skills away from the shop-floor; rigid controls versus delegation of decision-making over production.

The realities of this range of choice have been indicated by the work of Ian Nicholas and his colleagues in their comparison of the introduction of NC machine tools in Britain and Germany, in the work of Jones and the recent research of Wilkinson(27). But such studies share one common feature, a characteristic impossible to escape, they are all time-bound. They are all sampling organisations/firms at a particular historical point, one in which the form of the technology has not been closed off be a series of decisions and technical developments which in combination constitute sunk costs such that unwinding them, making a series of different choices, becomes an impossible cost burden. In other words, a series of short-term optimisations, enforced by market competition, results in an irrevocable situation(28).

But it is not just market pressures that determine technological choices, it is also historically-specific managerial ideologies and assumptions, such as those of Taylorism and later Fordism. Both these managerial philosophies assumed job simplification associated with immigrant labour, managerial centralisation and authoritarian control relationships. As such they helped to structure the forms of 'new technology' associated with the installation and development of new machine tools in the 1880-1910 period, and the development of flow-line production and new transfer mechanisms during the 1910-1950s period, a structuring which cumulatively foreclosed other options. Ford's assembly-lines became the image of modern factory production and as such was carried across the political frontiers to the Soviet Union with similar results(29).

At present we may be witnessing the emergence of a new cluster of ideas, a new ideology which marks out the parameters of future technological choices. The elements of that ideology - ITism - are not yet in place and there is no standard-bearer. Probably the ideology will be built around the concept of a system and, unlike most previous managerial ideologies, it will be seen as universally applicable across all sectors, not only manufacturing but also in services.

Several theorists(30) have argued that the processes of technical change do not follow a path of smooth progression, but reveal periods of rapid change based on technological clustering followed by periods of slower change. At present (1983), we seem to be in a period of rapid change such that decisions about technological design will be cumulatively 'frozen in', affecting and shaping the work-lives of succeeding generations perhaps for a century.

REFERENCES

1. GOLDING, H. The Wonderful World of Machinery. London: Ward Lock, 1933.

2. Two important efforts to achieve some integration are:

LANDES, D. Prometheus Unbound: Technological Change and Industrial Development in Western Europe from 1750 to the Present. Cambridge: Cambridge University Press, 1969;
NOBLE, D. America By Design: Science, Technology, and the Rise of Corporate Capitalism, London: Oxford University Press, 1977.

3. HILL, S. Economic and industrial transformation: the waves of social consequence from technological change. in: BOREHAM, P and DOW, G. (eds) Work and Inequality, Sydney: Macmillan, 1980.

4. BRIGHT, J.R. Automation and Management. Boston: Harvard University,1958.

5. BELL, M. Changing Technology and Manpower Requirements in the Engineering Industry. Brighton: Sussex University Press, 1972.

6. COOMBS, R. Automation, Management Strategies and Labour-Process Change. mimeo, 1983.

7. ibid, pp.14-24.

8. Adapted from COOMBS, op. cit., p.22.

9. I owe the term 'ITism' to Graham Winch. All blame rests with him!

10.CHAPMAN, S.D. The Textile Factory Before Arkwright: a Typology of Factory Development. in: Business History Review. 48, 1974.

11. ibid, p.471.

12. ibid, p.473.

13. See LITTLER, C.R.The Development of the Labour Process in Capitalist Societies: A Comparative Study of the Transformation of Work Organization in Britain, Japan and the USA. London: Heinemann, 1982.

14. HANNAH, L. The Rise of the Corporate Economy. London: Methuen, 1976.

15. ROSE, F.H. The Machine Monster: A Warning to Skilled Workers. London: ILP Publication Dept, 1909.

16. This does not represent a complete analysis of Taylorism. For such an analysis see LITTLER. op. cit., chapter 5.

17. ibid, and Littler, C.R. Taylorism, Fordism and Job Design. memeo, 1983.

18. FLINK, J.J. The Car Culture. Cambridge, Mass: MIT Press.

19. MEYER, S. The Five Dollar Day: Labor Management and Social Control in the Ford Motor Company 1908-21. Albany: State University of New York Press, 1981.

20. PORTER, H.F., Four big lessons from Ford's Factory. in: System, 31 June, 1917.

21. BRIGHT, op. cit., p. 15.

22. FORD, H. My Life and Work. Garden City, New York: Doubleday Page, 1922.

23. ARNOLD, H.L. Ford Methods and Ford Shops: Ford Motor-test Blocks and Chassis Assembly Lines. in: Engineering Magazine, 47, Aug. 1914.

24. LUND, R.T. Numerically Controlled Machine Tools and Group Technology: A Study of US Experience. Cambridge, Mass: MIT mimeo 1978.

15. TURNER, B. The Organization of Production-Scheduling in Complex Batch Production Situations. in HEALD, G. (ed). Approaches to the Study of Organizational Behaviour. London: Tavistock, 1970.

26. DIEBOLD, J. Automation: The Advent of the Automatic Factory. New York: Van Nostrand, 1952.

27. See JONES, B. Destruction or Redistribution of Engineering Skills? The Case of Numerical Control, in: WOOD, S. (ed) The Degradation of Work? London: Hutchinson, 1982; and WILKINSON, B. The Shop-Floor Politics of New Technology. London: Heinemann, 1983.

28. See Council for Science and Society. New Technology: Society, Employment and Skill. London: CSS, 1981.

29. LOWIT, T. Taylorisme et Controle Social en Europe de l'Est. Mimeo, 1983.

30. See MENSCH, G. Stalemate in Technology; Innovations Overcome The Depression, New York: Ballinger, 1979; and FREEMAN, C., CLARK, J., and SOETE, L. Unemployment and Technical Innovation. London: Frances Pinter, 1982.

NOTES ON CONTRIBUTORS

ERIK ARNOLD
Science Policy Research Unit,
University of Sussex.

JOHN BESSANT
Department of Business Studies,
Brighton Polytechnic.

DAVID BUCHANAN
Department of Management Studies,
University of Glasgow.

CYNTHIA COCKBURN
Department of Social Science and Humanities,
The City University,
London.

DAVID DAVIDSON
Centre for Employment Policy Studies,
The Management College, Henley.

JAMES FLECK
Technology Policy Unit,
University of Aston in Birmingham.

GERT HARTMANN
International Institute of Management,
West Berlin.

CRAIG R. LITTLER
Department of Social and Economic Studies,
Imperial College,
University of London.

PETER MURRAY
Department of Sociology,
Trinity College, Dublin.

IAN NICHOLAS
Ashridge Management College.

SHEILA ROTHWELL
Centre for Employment Policy Studies,
The Management College, Henley.

ARNDT SORGE
International Institute of Management,
West Berlin.

MALCOLM WARNER
The Management College, Henley.

JAMES WICKHAM
Department of Sociology,
Trinity College, Dublin.

GRAHAM WINCH
Department of Social and Economic Studies,
Imperial College,
University of London.